Instructor's Resource Manual
for Giancoli's

P H Y S I C S

Principles with Applications

Leroy W. Dubeck

Instructor's Resource Manual for Giancoli's

P H Y S I C S

Principles with Applications

THIRD EDITION

Douglas C. Giancoli

PRENTICE HALL, ENGLEWOOD CLIFFS, NEW JERSEY 07632

© 1991 by **PRENTICE-HALL, INC.**
A Simon & Schuster Company
Englewood Cliffs, N.J. 07632

10 9 8 7 6 5 4 3 2 1

ISBN 0-13-672882-0
Printed in the United States of America

CONTENTS

Introduction to Instructor's Resource Manual   v

Chapter 1   Introduction   1

Chapter 2   Describing Motion: Kinematics in One Dimension   2

Chapter 3   Kinematics in Two or Three Dimensions: Vectors   4

Chapter 4   Motion and Force: Dynamics   5

Chapter 5   Circular Motion: Gravitation   7

Chapter 6   Work and Energy   10

Chapter 7   Linear Momentum   12

Chapter 8   Rotational Motion   14

Chapter 9   Bodies in Equilibrium: Elasticity and Fracture   16

Chapter 10   Fluids   18

Chapter 11   Vibrations and Waves   21

Chapter 12   Sound   23

Chapter 13   Temperature and Kinetic Theory   26

Chapter 14   Heat   28

Chapter 15   The First and Second Laws of Thermodynamics   30

Chapter 16   Electric Charge and Electric Field   33

Chapter 17   Electric Potential and Electric Energy   35

Chapter 18   Electric Currents   37

Chapter 19   DC Currents and Instruments   39

Chapter 20   Magnetism   41

Chapter 21   Electromagnetic Induction and Faraday's Law;
             AC Circuits   44

Chapter 22   Electromagnetic Waves   46

Chapter 23   Light: Geometric Optics   49

CONTENTS

Chapter 24   The Wave Nature of Light   51

Chapter 25   Optical Instruments   54

Chapter 26   Special Theory of Relativity   56

Chapter 27   Early Quantum Theory and Models of the Atom   59

Chapter 28   Quantum Mechanics of Atoms   61

Chapter 29   Molecules and Solids   64

Chapter 30   Nuclear Physics and Radioactivity   66

Chapter 31   Nuclear Energy: Effects and Uses of Radiation   68

Chapter 32   Elementary Particles   72

Chapter 33   Astrophysics and Cosmology   74

Appendix: ABC News Archives Video Tapes   76

Appendix: Physics You Can See Video Tapes   77

Appendix: The Mechanical Universe...and Beyond Video Tapes   78

Appendix: The New York Times Newspaper Articles   80

Appendix: Transparencies   82

INTRODUCTION TO INSTRUCTOR'S RESOURCE MANUAL

This Resource Manual was prepared to save an instructor time in developing and preparing class presentations. It is a guide to instructors in physics courses using <u>Physics: Principles With Applications, third edition</u> by Douglas Giancoli.

A standard format is used for each chapter. The title of each subsection is stated and concepts, definitions, and formulae given in that subsection of the textbook are outlined. Suggested examples from the subsection are also listed.

At the end of most chapters suggested video tapes are described. There are three sets of such video tapes: "The Mechanical Universe...and Beyond," "Physics You Can See," and "The ABC News Archives." Finally, many chapters conclude with one or more relevant science articles from the New York Times. Both the video tapes and the New York Times articles reference the chapter subsections for which they are relevant. The Resource Manual includes a related question or problem for each of the New York Times articles and the ABC News Archives video tapes.

The three sets of video tapes, the New York Times articles, and **transparencies** are listed by chapter in separate appendices.

The textbook, and therefore this Resource Manual, covers more material than many instructors will be able to use in a one-year introductory physics course. The outline of the textbook given in this Resource Manual should help the instructor to select those chapters and subsections **that** are most relevant to the goals of a given course.

Instructor's Resource Manual
for Giancoli's

PHYSICS

Principles with Applications

CHAPTER 1

INTRODUCTION

1.1 SCIENCE AND CREATIVITY
   A. Description of a scientific theory
   B. Testing of a theory
   C. Acceptance of a theory

1.2 PHYSICS AND ITS RELATION TO OTHER FIELDS
   A. Physics used in biology and architecture

1.3 MODELS, THEORIES, AND LAWS
   A. Definition of a scientific model
   B. Contrast with a theory
   C. Definition of a law of nature

1.4 MEASUREMENT AND UNCERTAINTY
   A. All measurements have an uncertainty
   B. Estimating uncertainties
   C. Definition of significant figures
   D. Number of significant figures resulting from multiplications
      and divisions
   E. Expressing numbers in scientific notation

1.5 UNITS, STANDARDS AND THE SI SYSTEM
   A. Definitions of the standard meter and other units of length
   B. Definition of the second
   C. Description of the SI system
   D. Definition of base quantities and derived quantities

1.6 ORDER OF MAGNITUDE: RAPID ESTIMATING
   A. How to make an order of magnitude estimate of a quantity

1.7 MATHEMATICS
   A. Refer to mathematical concepts contained in the appendices

VIDEO TAPE: "INTRODUCTION TO THE MECHANICAL UNIVERSE" from The
Mechanical Universe...And Beyond introduces ideas and
scientists from Copernicus to Newton, and links physics on
earth to the heavens
For use with Sections 1.1 - 1.3

CHAPTER 2

DESCRIBING MOTION: KINEMATICS IN ONE DIMENSION

Definition of mechanics, kinematics, dynamics, and rectilinear motion

2.1 SPEED
    A. Definition of speed
    B. Example 2-1

2.2 REFERENCE FRAMES AND COORDINATE SYSTEMS
    A. Description of a frame of reference
    B. Value of a physical quantity depends upon the frame of reference
    C. Coordinate axes

2.3 CHANGING UNITS
    A. Converting units by multiplying by a ratio equal to one

2.4 AVERAGE VELOCITY AND DISPLACEMENT
    A. Definition of velocity
    B. Definition of displacement
    C. Example 2-2

2.5 INSTANTANEOUS VELOCITY
    A. Definition of instantaneous velocity
    B. Relationship of instantaneous to average velocity

2.6 VECTORS AND SCALARS
    A. Definition of vector
    B. Definition of scalar

2.7 ACCELERATION
    A. Definition of average acceleration
    B. Definition of instantaneous acceleration
    C. Example 2-3
    D. Example 2-4
    E. Decelerating means speed is decreasing

2.8 UNIFORMLY ACCELERATED MOTION
    A. Derivation of 4 kinematic equations (2-10a through 2-10d) when the acceleration is constant
    B. Example 2-5

2.9 SOLVING PROBLEMS
    A. Read problem carefully
    B. Write down the given quantities and those you wish to find
    C. Find applicable equation(s) and solve for the unknowns
    D. Do the solutions make sense
    E. Check the units on both sides of the equation(s) used to make sure they are equal

F. Example 2-7

2.10 FALLING BODIES
   A. Galileo postulated that all bodies on the earth fall with the same uniform acceleration in the absence of air friction
   B. The value of the acceleration due to gravity at the earth's surface
   C. Example 2-10
   D. Example 2-11

2.11 GRAPHICAL ANALYSIS OF LINEAR MOTION
   A. Graphs of position versus time
   B. Graphs of velocity versus time
   C. Relationship between displacement and area under velocity versus time graphs
   D. Example 2-13

   VIDEO TAPE: "VECTORS" from the Mechanical Universe... And Beyond describes the importance of vectors to physics For use with Section 2.6

   VIDEO TAPE: "THE LAW OF FALLING BODIES" from the Mechanical Universe... and Beyond describes Galileo's experiments which proved that all bodies fall with the same acceleration at the earth's surface if air resistance can be neglected For use with Section 2.10

CHAPTER 3

KINEMATICS IN TWO OR THREE DIMENSIONS: VECTORS

3.1 ADDITION OF VECTORS--GRAPHICAL METHODS
   A. Addition of 2 vectors along the same line
   B. Addition of 2 vectors at right angles to one another
   C. Tail to tip method of adding any two vectors graphically
   D. Parallelogram method of adding any two vectors

3.2 SUBTRACTION OF VECTORS AND MULTIPLICATION OF VECTOR BY A SCALAR
   A. Subtraction of one vector from another
   B. Multiplication of a vector by a scalar

3.3 ANALYTIC METHOD FOR ADDING VECTORS: COMPONENTS
   A. Resolving a vector into its components
   B. Discussion of the sine, cosine and tangent functions
   C. Using these functions to resolve a vector into its
      components
   D. Addition of vectors by adding their components to find the
      components of the resultant vector
   E. Example 3-1

3.4 RELATIVE VELOCITY - VECTORS IN PROBLEM SOLVING
   A. Each velocity labelled by 2 subscripts: the first refers to
      the object and the second to the reference frame in which
      it has this velocity
   B. Addition of vectors using this subscript notation
   C. Example 3-2
   D. Example 3-3

3.5 PROJECTILE MOTION
   A. Definition of projectile motion
   B. Vertical displacement formula
   C. Horizontal velocity constant
   D. An object projected horizontally reaches the ground at the
      same time as an object dropped vertically from that height

3.6 SOLVING PROBLEMS INVOLVING PROJECTILE MOTION
   A. One applies Equations 2-10a through 2-10d
   B. Example 3-6
   C. Example 3-8

VIDEO TAPE: PHYSICS YOU CAN SEE: "MONKEY AND GUN" demonstrates
the rate of vertical free fall
For use with Section 3.5
Running Time: 2:07

CHAPTER 4

MOTION AND FORCE: DYNAMICS

4.1 FORCE
   A. Definition of force
   B. Force is a vector

4.2 NEWTON'S FIRST LAW OF MOTION
   A. Contrast Aristotle's and Galileo's views of motion
   B. Statement of Newton's First Law of Motion
   C. Definition of inertia

4.3 MASS
   A. Definition of mass
   B. Standard units of mass: the kilogram and the atomic mass
      unit

4.4 NEWTON'S SECOND LAW OF MOTION
   A. Statement of Newton's Second Law of Motion
   B. Units of force: the newton, the dyne and the pound
   C. Example 4-1

4.5 LAWS OR DEFINITIONS
   A. Difference between a law and a definition in physics
   B. Inertial reference frame
   C. Noninertial reference frame

4.6 NEWTON'S THIRD LAW OF MOTION
   A. Statement of Newton's Third Law of Motion
   B. Examples of this law: pushing a desk, throwing a package out
      of a boat, firing a rocket, moving an automobile
   C. Example 4-3

4.7 WEIGHT - THE FORCE OF GRAVITY; AND THE NORMAL FORCE
   A. Statement that the weight of a body equals its mass times
      the acceleration due to gravity
   B. Discussion of the acceleration due to gravity near the
      surface of the earth
   C. Definition and discussion of the normal force
   D. The action and reaction forces of Newton's Third Law act on
      different objects
   E. Example 4-4

4.8 SOLVING PROBLEMS WITH NEWTON'S LAWS: VECTOR FORCES
   A. Net force is the vector sum of all forces acting on object
   B. Example 4-5
   C. Sketching a free-body diagram
   D. Example 4-7
   E. Example 4-9

4.9 PROBLEMS INVOLVING FRICTION, INCLINES

5

A. Definition of kinetic friction and its relationship to the normal force between surfaces
B. Definition of static friction
C. Coefficients of kinetic and static friction
D. Example 4-11
E. Example 4-13
F. Normal and frictional forces on an inclined plane
G. Example 4-14
H. Example 4-16

4.10 PROBLEM SOLVING - A GENERAL APPROACH
A. Read problem carefully
B. Draw free-body diagram
C. Choose convenient x-y coordinate system
D. Determine what the unknowns are
E. Determine which equations are applicable to the problem
F. Try to solve the problem roughly (see Section 1.6)
G. Solve the problem: keep track of the units used
H. Recheck to see if the answer is reasonable

VIDEO TAPE: "INERTIA" from The Mechanical Universe...And Beyond discusses Galileo's views of inertia
For use with Section 4.2

VIDEO TAPE: "NEWTON'S LAWS" from The Mechanical Universe...And Beyond describes Newton's statements of his Laws of Motion
For use with Sections 4.2 - 4.6

VIDEO TAPE: PHYSICS YOU CAN SEE: "COIN AND FEATHER" demonstrates the acceleration due to gravity
For use with Section 4.7
Running Time: 1:30

NEW YORK TIMES ARTICLE: "PHYSICS AND COMPUTERS CREATE A SCIENCE OF AQUATIC THRILLS" describes the designs of water slides and artificial water waves for surfing
For use with Section 4.9

Additional problem: Calculate the speed of a rider at the bottom of an aquatic slide that is 60 feet high and at an angle of 55.5 degrees above the horizontal. Neglect friction and assume that the rider starts from rest at the top of the slide. (Answer: 62 feet/second or 42 miles per hour)

CHAPTER 5

CIRCULAR MOTION: GRAVITATION

5.1 KINEMATICS OF UNIFORM CIRCULAR MOTION
   A. Definition of uniform circular motion
   B. Derivation of the formula for the centripetal acceleration of an object moving in a circle at constant speed
   C. Example 5-1
   D. Example 5-2

5.2 DYNAMICS OF UNIFORM CIRCULAR MOTION
   A. Centripetal force not some new type of force
   B. Centrifugal force does not exist
   C. Example 5-3

5.3 NONUNIFORM CIRCULAR MOTION
   A. The tangential component of acceleration
   B. The total acceleration is the vector sum of the centripetal and tangential accelerations

5.4 CENTRIFUGATION
   A. Description of a centrifuge
   B. Example 5-6

5.5 NEWTON'S LAW OF UNIVERSAL GRAVITATION
   A. Newton's derivation of his law of universal gravitation
   B. Statement of the law of universal gravitation
   C. The value of the universal gravitational constant
   D. Example 5-7

5.6 GRAVITY NEAR THE EARTH'S SURFACE: GEOPHYSICAL APPLICATIONS
   A. The derivation of the acceleration due to gravity at the surface of the earth
   B. Example 5-9

5.7 SATELLITES AND WEIGHTLESSNESS
   A. The relationship between the speed and the orbital radius of a satellite
   B. Example 5-10
   C. Apparent weightlessness in a satellite and in an elevator
   D. Effects on human beings of weightlessness

5.8 EARTH'S TIDES
   A. Explanation of why there are two high and low tides every 25 hours

5.9 KEPLER'S LAWS AND NEWTON'S SYNTHESIS
   A. Statements of Kepler's three laws of planetary motion
   B. The derivation of Kepler's third law of planetary motion
   C. Example 5-12
   D. Example 5-13

E. Perturbations of the orbits of planets from ellipses led to the discovery of additional planets whose gravitational attractions were responsible for the perturbations

5.10 TYPES OF FORCES IN NATURE
    A. The gravitational force
    B. The electromagnetic force
    C. The strong nuclear force
    D. The weak nuclear force

VIDEO TAPE: "MOVING IN CIRCLES" from The Mechanical Universe ...And Beyond looks at the physics of uniform circular motion and Newton's contributions to that study
For use with Section 5.2

VIDEO TAPE: "THE APPLE AND THE MOON" From The Mechanical Universe...And Beyond describes Newton's discovery of the force between any two particles in the universe
For use with Section 5.5

VIDEO TAPE: "KEPLER'S THREE LAWS" From The Mechanical Universe...And Beyond describes these laws
For use with Section 5.9

VIDEO TAPE: "THE KEPLER PROBLEM" From The Mechanical Universe...And Beyond deduces Kepler's laws from Newton's law of universal gravitation
For use with Section 5.9

VIDEO TAPE: "FUNDAMENTAL FORCES" From The Mechanical Universe...And Beyond describes the 4 fundamental forces of nature
For use with Section 5.10

VIDEO TAPE: "WALKING IN SPACE" from ABC News archives describes the health problems experienced by astronauts due to being in a "weightless" condition
For use with Section 5.7
How could astronauts overcome these health problems in a future large scale space station?
(Answer: having a circular shaped space station rotate to create an artificial gravity at its outer rim approximately equal to that on the earth's surface)
Running Time: 3:16

VIDEO TAPE: PHYSICS YOU CAN SEE: "TEST TUBE ON A WHEEL" demonstrates centripetal acceleration
For use with Section 5.1
Running Time: 2:40

VIDEO TAPE: PHYSICS YOU CAN SEE: "WATER WHIRLED IN A CIRCLE" demonstrates centripetal force

8

For use with Section 5.2
Running Time: 1:05

NEW YORK TIMES ARTICLE: "FINNISH SCIENTISTS TO STUDY GRAVITY IN ECLIPSE" describes an experiment to test whether the gravity of one massive body (the moon) can partly shield that of another (the sun)
For use with Section 5.5

Additional problem: Suppose the moon reduces the gravitational acceleration at the earth's surface by 10^{-5} meters/sec^2. How much longer would an object take to fall 4.9 meters from rest? (Answer: 5×10^{-6} sec)

CHAPTER 6

WORK AND ENERGY

6.1 WORK DONE BY A CONSTANT FORCE
   A. Definition of work done by a constant force acting on an
      object
   B. Units for work
   C. A force can be present without doing work
   D. Example 6-1
   E. Example 6-2

6.2 WORK DONE BY A VARYING FORCE
   A. Graphical determination of the work done by a varying force

6.3 KINETIC ENERGY AND THE WORK-ENERGY THEOREM
   A. Definition of energy as the ability to do work
   B. Definition of kinetic energy and the derivation of the
      formula for kinetic energy
   C. Statement of the work-energy theorem
   D. Energy units
   E. Example 6-4
   F. Example 6-5

6.4 POTENTIAL ENERGY
   A. Definition of potential energy
   B. Gravitational potential energy
   C. Example 6-6
   D. Formula for the change in gravitational potential energy
   E. General relationship between the change in potential energy
      and the force producing that change
   E. Formula for the change in elastic potential energy

6.5 CONSERVATIVE FORCES
   A. Definition of a conservative force
   B. Potential energy can only be defined for a conservative
      force
   C. General form of the work-energy theorem

6.6 OTHER FORMS OF ENERGY: ENERGY TRANSFORMATIONS
   A. Other types of energy
   B. Work is done whenever energy is transformed from one object
      to another

6.7 THE LAW OF CONSERVATION OF ENERGY
   A. Statement of the law of conservation of energy
   B. Conditions under which the sum of the kinetic and potential
      energies of a system is conserved

6.8 PROBLEM SOLVING USING CONSERVATION OF MECHANICAL ENERGY
   A. Discussion of objects moving under gravity without friction
   B. Example 6-7

6.9 ENERGY CONSERVATION WITH DISSIPATIVE FORCES: SOLVING PROBLEMS
   A. Definition of dissipative forces
   B. Conservation of energy with gravity and friction
   C. Example 6-10

6.10 POWER
   A. Definition of power
   B. Units of power
   C. Example 6-11
   D. Example 6-12

   VIDEO TAPE: "POTENTIAL ENERGY" from The Mechanical
   Universe...And Beyond uses potential energy as a way to
   understand the world around us
   For use with Section 6.4

   VIDEO TAPE: "CONSERVATION OF ENERGY" from The Mechanical
   Universe...And Beyond discusses this law of nature
   For use with Section 6-7

CHAPTER 7

LINEAR MOMENTUM

7.1 MOMENTUM AND ITS RELATION TO FORCE
  A. Definition of linear momentum
  B. Restatement of Newton's second law of motion in terms of momentum
  C. Example 7-1

7.2 CONSERVATION OF MOMENTUM
  A. Derivation of the conservation of momentum theorem for a one dimensional collision
  B. Statement of the law of conservation of momentum
  C. Example 7-2
  D. Example 7-3

7.3 COLLISIONS AND IMPULSE
  A. Definition of impulse
  B. Example 7-4

7.4 CONSERVATION OF ENERGY AND MOMENTUM IN COLLISIONS
  A. Definition of elastic collisions
  B. Definition of inelastic collisions

7.5 ELASTIC COLLISIONS IN ONE DIMENSION
  A. Momentum and kinetic energy are both conserved
  B. Example 7-5

7.6 ELASTIC COLLISIONS IN TWO OR THREE DIMENSIONS
  A. Momentum and kinetic energy are again conserved
  B. Example 7-6

7.7 INELASTIC COLLISIONS
  A. Definition of a completely inelastic collision
  B. Example 7-7
  C. Example 7-8

7.8 CENTER OF MASS
  A. Definition and formula for the center of mass of an extended body
  B. The center of mass as a means of describing the motion of an extended object or group of objects
  C. Example 7-9

7.9 CENTER OF MASS AND TRANSLATIONAL MOTION
  A. Description of the motion of a system of 3 bodies in one dimension using the center of mass
  B. Statement of Newton's second law for a system of particles
  C. Example 7-11

VIDEO TAPE: "CONSERVATION OF MOMENTUM" from The Mechanical Universe...And Beyond discusses the conservation of momentum principle
For use with Section 7.2

VIDEO TAPE: PHYSICS YOU CAN SEE: "SWIVEL HIPS" demonstrates force pairs
For use with Section 7.2
Running Time: 1:24

CHAPTER 8

ROTATIONAL MOTION

8.1 ANGULAR QUANTITIES
   A. Definition of the radian
   B. Example 8-1
   C. Definition of angular velocity
   D. Definition of angular acceleration
   E. Relationship of tangential velocity to angular velocity
   F. Relationship of tangential linear acceleration to angular acceleration
   G. Expressing centripetal acceleration in terms of angular quantities
   H. Definition of frequency
   I. Definition of the period
   J. Example 8-2
   K. Example 8-3
   L. Example 8-4

8.2 KINEMATIC EQUATIONS FOR UNIFORMLY ACCELERATED ROTATIONAL MOTION
   A. The statement of 4 rotational kinematic equations (8-9a through 8-9d) equivalent to the linear kinematic equations
   B. Example 8-5

8.3 TORQUE
   A. Definition of the lever arm of a force about an axis of rotation
   B. Definition of the torque of a given force about an axis of rotation
   C. Example 8-6
   D. Assigning signs to torques

8.4 ROTATIONAL DYNAMICS: TORQUE AND ROTATIONAL INERTIA
   A. Angular acceleration is proportional to the applied torque
   B. Definition of the moment of inertia
   C. Moment of inertia plays same role for rotational motion as mass does for translational motion

8.5 SOLVING PROBLEMS IN ROTATIONAL DYNAMICS
   A. Example 8-8
   B. Definition of radius of gyration
   C. Example 8-9

8.6 ROTATIONAL KINETIC ENERGY
   A. Formula for rotational kinetic energy
   B. The total kinetic energy of an object equals the kinetic energy of its center of mass plus its rotational kinetic energy about its center of mass
   C. Example 8-11

8.7 ANGULAR MOMENTUM AND ITS CONSERVATION
   A. Definition of angular momentum
   B. Rotational equivalent of Newton's second law
   C. Statement of the law of conservation of angular momentum
   D. Applications to the motions of skaters and divers
   E. Example 8-12

8.8 VECTOR NATURE OF ANGULAR QUANTITIES
   A. The right hand rule for the directions of the angular
      velocities and the torques

8.9 VECTOR ANGULAR MOMENTUM: A ROTATING WHEEL
   A. Discussion of the motion of a person walking on a freely
      rotating platform
   B. Discussion of tilting a spinning bicycle wheel
   C. Description of the operation of a gyroscope

8.10 ROTATING FRAMES OF REFERENCE: INERTIAL FORCES
   A. Definition of inertial and non-inertial frames of reference
   B. Pseudoforces in non-inertial reference frames
   C. Application to a centrifuge

8.11 THE CORIOLIS FORCE
   A. Description of the origin of the Coriolis acceleration
   B. Derivation of the Coriolis acceleration for a rotating
      platform
   C. Application of the Coriolis acceleration to air mass motion
      on the earth

   VIDEO TAPE: "ANGULAR MOMENTUM" from The Mechanical
   Universe...And Beyond describes this concept
   For use with Section 8.7

   VIDEO TAPE: "TORQUES AND GYROSCOPES" from The Mechanical
   Universe...And Beyond describes torques and applies angular
   momentum and torque to everything from spinning tops to the
   equinoxes
   For use with Sections 8.9, 8.10, and 8.11

CHAPTER 9

BODIES IN EQUILIBRIUM: ELASTICITY AND FRACTURE

9.1 STATICS--THE STUDY OF FORCES IN EQUILIBRIUM
    A. Definition of an object in equilibrium
    B. Example 9-1

9.2 THE CONDITIONS FOR EQUILIBRIUM
    A. The sum of the components of the forces acting on a body in both the x direction and the y direction is zero
    B. Example 9-2
    C. The sum of the torques acting on a body must also be zero

9.3 SOLVING STATICS PROBLEMS
    A. Choose one body at a time and make a free-body diagram of it
    B. Choose a coordinate system and resolve all forces into their components
    C. Write down the three equilibrium equations and solve
    D. Example 9-3
    E. Example 9-4
    F. Example 9-6

9.4 APPLICATIONS TO MUSCLES AND JOINTS
    A. Definitions of insertions, joints, flexor and extensor muscles
    B. Example 9-7
    C. Discussion of forces on the human spinal column
    D. Example 9-8

9.5 SIMPLE MACHINES: LEVERS, PULLEYS, AND MULTISPEED BICYCLES
    A. Definition of ideal mechanical advantage
    B. Definition of actual mechanical advantage
    C. Definition of efficiency of a machine
    D. Application to a pulley
    E. Application to a multi-speed bicycle

9.6 STABILITY AND BALANCE
    A. Definitions of stable, unstable, and neutral equilibriums
    B. Relationship between center of gravity, base of support, and stability
    C. Humans walk to adjust center of gravity to base of support

9.7 ELASTICITY: STRESS AND STRAIN
    A. Statement of Hooke's law
    B. Definition of proportional and elastic limits
    C. Definition of Young's modulus
    D. Definition of stress and strain
    E. Example 9-9
    F. Definition of compressive stress
    G. Definition of shear stress

16

H. Definition of shear modulus
I. Definition of bulk modulus

9.8 FRACTURE
   A. Structures need to have factors of 3 to 10 or more safety margins against fractures due to stresses
   B. Example 9-10
   C. Definitions of reinforced and prestressed concretes
   D. Example 9-11

9.9 SPANNING A SPACE: ARCHES AND DOMES
   A. Semicircular arches allow their composite stones to experience mainly compressive forces which they can withstand much better than tension forces
   B. Advantage of using a pointed arch
   C. Force analysis of domes
   D. Example 9-12

   VIDEO TAPE: PHYSICS YOU CAN SEE: "EQUILIBRIUM" demonstrates isolation of forces
   For use with Section 9.2
   Running Time: 3:47

CHAPTER 10

FLUIDS

The three common states of matter are the solid, liquid and gas:
each is defined.
Fluid is defined.

10.1 DENSITY AND SPECIFIC GRAVITY
     A. Definition and units of density
     B. Example 10-1

10.2 PRESSURE IN FLUIDS
     A. Definition and units for pressure
     B. A fluid exerts a pressure in all directions, perpendicular
        to any container walls
     C. Formula for the pressure exerted by a liquid
     D. Example 10-2

10.3 ATMOSPHERIC PRESSURE AND GAUGE PRESSURE
     A. The value and units of atmospheric pressure
     B. The definition of gauge pressure

10.4 PASCAL'S PRINCIPLE
     A. Statement of Pascal's principle
     B. Application to a hydraulic lift

10.5 MEASUREMENT OF PRESSURE: GAUGES AND THE BAROMETER
     A. Definition of the Torr and mm of Hg as units of pressure
     B. Description of a Bourdon gauge and a mercury barometer
     C. Water can be raised no more than 10.3 meters by a vacuum
        pump

10.6 BUOYANCY AND ARCHIMEDES PRINCIPLE
     A. Definition of the buoyant force
     B. Statement of Archimedes principle
     C. Example 10-3
     D. Example 10-4
     E. Example 10-5

10.7 FLUIDS IN MOTION: FLOW RATE AND THE EQUATION OF CONTINUITY
     A. Definition of hydrodynamics
     B. Description of streamline flow and turbulent flow
     C. Derivation of the equation of continuity
     D. Example 10-8

10.8 BERNOULLI'S EQUATION
     A. Derivation of Bernoulli's equation

10.9 APPLICATIONS OF BERNOULLI'S PRINCIPLE: FROM TORRICELLI TO
     SAILBOATS, AIRFOILS AND TIA
     A. Torricelli's theorem

B. How airplanes fly
C. How sailboats function
D. Description of a Venturi tube
E. Why smoke rises up a chimney
F. Explanation of "transient ischemic attack"
G. Example 10-10

10.10 VISCOSITY
A. Definition of viscosity
B. The coefficient of viscosity

10.11 FLOW IN TUBES: POISEUILLE'S EQUATION, BLOOD FLOW, REYNOLDS NUMBER
A. Statement of Poiseuille's equation
B. Example 10-11
C. Definition of the Reynolds number
D. Application to blood flow in the human body
E. Example 10-12

10.12 OBJECT MOVING IN A FLUID: SEDIMENTATION AND DRAG
A. Definition of a second Reynolds number
B. Statement of Stoke's equation
C. Definition of sedimentation
D. Determination of an object's terminal velocity

10.13 SURFACE TENSION AND CAPILLARITY
A. Definition of surface tension
B. Examination from a molecular viewpoint
C. Example 10-14
D. Description of adhesion and cohesion
E. Description of capillary action
F. Example 10-15

10.14 NEGATIVE PRESSURE AND THE COHESION OF WATER: THE RISE OF FLUIDS IN TREES
A. An apparatus to produce negative pressures is described
B. This phenomenon is believed to account for the rise of water in tall trees

10.15 PUMPS: THE HEART AND BLOOD PRESSURE
A. Vacuum pumps and force pumps are described
B. A circulating pump, the heart, is described
C. Systolic and diastolic blood pressures are defined and their measurements are described

VIDEO TAPE: PHYSICS YOU CAN SEE: "COLLAPSE A CAN"
demonstrates atmospheric pressure
For use with Section 10.3
Running Time: 1:58

NEW YORK TIMES ARTICLE: "SAILS AND WIND SPEED" describes how
a sailing vessel can move faster than the actual speed of the

wind
For use with Section 10.9

Additional problem: How can a sailboat move against the wind?
(Answer: see page 253 and figure 10-19)

NEW YORK TIMES ARTICLE: "AIRPLANE SPEED" describes how an
airplane uses a pitot tube to measure air speed
For use with Section 10.9

Additional problem: What equation is used to determine the
local air speed at the entrance to the pitot tube? (Answer:
Bernoulli's equation)

**NEW YORK TIMES ARTICLE: "NEW PLANE WING DESIGN GREATLY CUTS
DRAG TO SAVE FUEL"** describes how covering a wing with a
porous metal skin changes turbulent air flow to laminar air
flow
For use with Section 10.9

Additional problem: Why is it necessary to have a Krueger
flap as part of the wing for the system to work? (Answer: the
flap serves as a shield, when extended forward, against
collisions with insects whose bodies would spoil the laminar
flow of air and would also clog the vital suction holes in
the new wing construction)

20

CHAPTER 11
VIBRATIONS AND WAVES

11.1 SIMPLE HARMONIC MOTION
  A. Definitions of vibrations, periodic motion, equilibrium position, displacement, amplitude, period, and frequency
  B. Description of the motion of an oscillating spring
  C. Definition of simple harmonic motion
  D. Example 11-1

11.2 ENERGY IN THE SIMPLE HARMONIC OSCILLATOR
  A. Expression for the total mechanical energy of an oscillating spring
  B. Example 11.2

11.3 THE REFERENCE CIRCLE: THE PERIOD AND SINUSOIDAL NATURE OF SHM
  A. Expressions are derived for the speed and the period of an object undergoing SHM
  B. Example 11-4

11.4 THE SIMPLE PENDULUM
  A. Derivation of the period of a simple pendulum
  B. Example 11-7

11.5 DAMPED HARMONIC MOTION
  A. Definition of damped harmonic motion
  B. Overdamped, underdamped, and critical damping are described

11.6 FORCED VIBRATIONS: RESONANCE
  A. The natural frequency of a body is defined
  B. Forced vibration is defined
  C. Resonance and the resonant frequency are defined
  D. Examples of resonance are described

11.7 WAVE MOTION
  A. A wave pulse and a periodic wave are defined
  B. If the source of a wave exhibits SHM, then so will the wave
  C. A periodic sinusoidal wave is described in terms of its amplitude, wavelength, and frequency
  D. The wave velocity is defined and expressions for it are given
  E. Example 11-8

11.8 TYPES OF WAVES
  A. Transverse and longitudinal waves are defined
  B. Examples are given for each type of wave
  C. Only longitudinal waves can pass through a liquid
  D. Example 11-9

11.9 ENERGY TRANSMITTED BY WAVES
  A. The energy transported by a wave is equal to the square of

its amplitude
B. The intensity of a wave is also proportional to the square of its amplitude
C. Example 11-10

11.10 BEHAVIOR OF WAVES: REFLECTION, REFRACTION, INTERFERENCE AND DIFFRACTION
A. In the reflection of waves, the angle of incidence equals the angle of reflection
B. Refraction is defined and the law of refraction is stated
C. Example 11-11
D. Interference between waves is described
E. The principle of superposition is stated
F. The phase of a wave is defined
G. Diffraction is defined

11.11 STANDING WAVES: RESONANCE
A. Definition of standing wave
B. The resonant frequencies of a rope fixed at both ends: the fundamental frequency and the harmonics
C. Example 11-12

VIDEO TAPE: "HARMONIC MOTION" from The Mechanical Universe...And Beyond describes the mathematics of periodic motion
For use with Section 11.1

VIDEO TAPE: "WAVES" from The Mechanical Universe ... And Beyond describes wave motion and the propagation of sound
For use with Section 11.7

VIDEO TAPE: "RESONANCE" from The Mechanical Universe ... And Beyond describes the collapse of a swaying bridge with a high wind and the shattering of a glass by a singer
For use with Section 11.12

VIDEO TAPE: PHYSICS YOU CAN SEE: "SINGING PIPES AND SINGING ROD" demonstrates longitudinal standing waves
For use with Section 11.12
Running Time: 1:25

CHAPTER 12

SOUND

12.1 CHARACTERISTICS OF SOUND
    A. The speed of sound in air as a function of temperature is given
    B. The pitch of a sound is defined as its frequency and the audible range is stated
    C. Ultrasonic and infrasonic sound waves are defined

12.2 INTENSITY OF SOUND
    A. Loudness is defined as the intensity of the wave
    B. The intensity level of any sound is defined in terms of the decibel
    C. Sample calculations using the decibel are given
    D. Example 12-1

12.3 INTENSITY RELATED TO AMPLITUDE AND PRESSURE AMPLITUDE
    A. The intensity of a wave is proportional to the square of the wave amplitude
    B. Sound waves are longitudinal waves, often called pressure waves
    C. The intensity is proportional to the square of the pressure amplitude and decreases as the inverse square of the distance from the sound source
    D. Example 12-3

12.4 THE EAR AND ITS RESPONSE; LOUDNESS
    A. The outer ear, the middle ear, and the inner ear are described
    B. The ear's sensitivity to the frequency of the incoming sound is described and graphs illustrate the sensitivity of the human ear
    C. The loudness level in units called phons is defined

12.5 SOURCES OF SOUND: VIBRATING STRINGS AND AIR COLUMNS
    A. The origin of sound in musical instruments is described
    B. The resonant frequencies in stringed instruments are described
    C. Example 12-4
    D. Standing waves are described for a tube open at both ends and for a tube closed at one end
    E. Overtones are defined
    F. Example 12-5

12.6 QUALITY OF SOUND, AND NOISE
    A. The quality of a sound depends on the presence of overtones
    B. The effects of noise on humans and the means of controlling noise are discussed

12.7 INTERFERENCE OF SOUND WAVES: BEATS

A. The conditions for constructive interference and the destructive interference between two sound waves are described
B. Example 12-8
C. Beats are defined and the formula for the beats produced by two sound sources is given

12.8 THE DOPPLER EFFECT
A. The period of a wave is stated
B. The equations relating frequencies are derived for relative motions between source and observer
C. Example 12-9

12.9 SHOCK WAVES AND THE SONIC BOOM
A. The Mach is defined
B. A shock wave is defined and described
C. The sonic boom produced by a shock wave is discussed

12.10 APPLICATION; ULTRASOUND AND MEDICAL IMAGING
A. The sonar or pulse echo technique is described
B. Sonar generally uses ultrasonic frequencies
C. In medicine the pulse echo technique is used much like sonar to locate abnormal growths
D. The frequencies used in ultrasonic diagnosis are given
E. The limitations of ultrasound imaging are discussed

VIDEOTAPE: "GOODBYE GALLSTONES" from ABC News archives describes the use of ultrasound to break up gallstones into pieces small enough to pass out through the ducts and then be eliminated through the bowel. The limitations of this procedure and its successes are discussed.
For use with Section 12.10
Ask students to research the current medical view about the recurrence of gallstones after treatment with lithotripsy. The video tape stated "We should expect that about 50% of patients after lithotripsy will have a recurrence of gallstones."
Running Time: 11:46

NEW YORK TIMES ARTICLE: "SOUND OVER WATER" describes the way sound travels over water
For use with Section 12.1

Additional problem: If the air over the water is cooler than that over the surrounding land, in what direction are sound waves bent in crossing the water? (Answer: towards the ground)

NEW YORK TIMES ARTICLE: "OCEAN LOUDSPEAKERS TO SOUND OFF FOR DATA ON GLOBAL WARNING" describes the use of a world girdling layer of water in the oceans in which sound travels relatively slowly. The speed of sound in this water layer will be measured to determine increases in water temperature.

24

For use with Section 12.1

Additional problem: If the travel time for sound going the 11,200 mile distance from Herd Island to San Francisco changes by .25 seconds in a year, what is the approximate percentage change in the speed of sound in water during that year? (Answer: .0022%)

NEW YORK TIMES ARTICLE: "DESIGNING AN SST: NOISE, SONIC BOOMS AND THE OZONE LAYER" describes the environmental problems that must be overcome in designing a new supersonic transport plane.

For use with Section 12.9

Additional problem: Can engine configuration and sound suppressing designs reduce noise substantially at take-off and landing? (Answer: we don't know at this time whether this can be done.)

CHAPTER 13

TEMPERATURE AND KINETIC THEORY

13.1 ATOMS
   A. The atom is defined
   B. The atomic mass is defined
   C. Brownian motion is described
   D. This chapter discusses phenomena from both a microscopic and macroscopic point of view

13.2 TEMPERATURE
   A. Temperature and thermometer are defined
   B. The Celsius and Fahrenheit temperature scales are described and a conversion formula given
   C. Example 13-1
   D. The constant-volume gas thermometer is described

13.3 THERMAL EXPANSION
   A. The coefficient of linear expansion is defined and the formula for linear expansion is given
   B. Example 13-2
   C. The coefficient of volume expansion and the change in volume with temperature are given
   D. Example 13-4

13.4 THERMAL STRESSES
   A. Thermal stresses are defined and a formula relating them to changes in temperature is given
   B. Example 13-5

13.5 THE GAS LAWS AND ABSOLUTE TEMPERATURE
   A. Boyle's law is given
   B. Charles' law is given
   C. Absolute zero is described and the Kelvin temperature scale is given
   D. Gay Lussac's law is given

13.6 THE IDEAL GAS LAW
   A. The prior three laws are combined into the ideal gas law and the universal gas constant is given
   B. Example 13-6
   C. Example 13-7

13.7 IDEAL GAS LAW IN TERMS OF MOLECULES: AVOGADRO'S NUMBER
   A. Avogadro's hypothesis and number are presented
   B. Boltzmann's constant is given
   C. Example 13-9

13.8 KINETIC THEORY AND THE MOLECULAR INTERPRETATION OF TEMPERATURE
   A. The microscopic properties of an ideal gas are stated
   B. The macroscopic pressure is related to the microscopic

kinetic energy of molecules
C. The average translational kinetic energy of molecules in a gas is shown to be directly proportional to the absolute temperature
D. Example 13-11
E. Example 13-13

13.9 DISTRIBUTION OF MOLECULAR SPEED
A. The Maxwell distribution of molecular speeds is presented graphically
B. The activation energy for a chemical reaction is defined

13.10 REAL GASES AND CHANGES OF PHASE
A. The critical temperature and the critical point are defined
B. The phase diagram of a substance is presented
C. Sublimation is defined
D. The triple point is defined
E. Superfluidity is defined

13.11 VAPOR PRESSURE AND HUMIDITY
A. Evaporation is defined and described
B. Condensation is defined and described
C. Saturated vapor pressure is defined and described
D. Boiling is defined and described
E. The total pressure is the sum of the partial pressures of each gas present
F. Relative humidity is defined
G. Example 13-14
H. Supersaturated air is defined
I. Dew point is defined

13.12 DIFFUSION
A. The diffusion constant is defined and the diffusion equation is stated
B. Example 13-15
C. The importance of diffusion for plants and animals is discussed

VIDEO TAPE: "TEMPERATURE AND GAS LAWS" from The Mechanical Universe...and Beyond describes the behavior of gases and the connection between temperature and heat.
For use with Sections 13.5 through 13.8

VIDEO TAPE: "LOW TEMPERATURES" from The Mechanical Universe ...and Beyond describes the quest for low temperatures and the discovery that all elements can exist in each of the basic states of matter.
For use with Section 13.5

CHAPTER 14

HEAT

14.1 HEAT AS ENERGY TRANSFER
   A. The calorie, the kilocalorie, and the British thermal unit
      are each defined
   B. The mechanical equivalent of heat is given
   C. Heat is defined as the energy transferred from one body to
      another because of a difference in temperature

14.2 DISTINCTION BETWEEN TEMPERATURE, HEAT, AND INTERNAL ENERGY
   A. The thermal energy or internal energy is defined

14.3 INTERNAL ENERGY OF AN IDEAL GAS
   A. The formula for the internal energy of an ideal gas is
      derived

14.4 SPECIFIC HEAT
   A. The formula relating the heat added to a system to its
      specific heat and rise in temperature is given
   B. Example 14-1

14.5 CALORIMETRY - SOLVING PROBLEMS
   A. For an isolated system the heat lost by one part of the
      system is equal to the heat gained by the other part
   B. Example 14-2
   C. A calorimeter is described
   D. Example 14-3

14.6 LATENT HEAT, AND PROBLEM SOLVING
   A. Changes of phase are described and the latent heat needed
      to produce these changes is defined and discussed
   B. Example 14-5
   C. Evaporation is described

14.7 HEAT TRANSFER: CONDUCTION
   A. Heat conduction is described: it takes place only if there
      is a difference in temperature
   B. The thermal conductivity constant is defined and related
      to other properties of the material
   C. Good conductors and good insulators are described
   D. Example 14-8

14.8 HEAT TRANSFER: CONVECTION
   A. Convection is defined and examples presented

14.9 HEAT TRANSFER: RADIATION
   A. The transfer of energy through radiation is defined and
      described
   B. The Stefan Boltzmann equation is presented and the Stefan
      Boltzmann constant is given

C. The emissivity is defined
D. A good heat absorber is also a good heat emitter
E. Example 14-9
F. Heat exchange between a human body and its surroundings is discussed

CHAPTER 15

THE FIRST AND SECOND LAWS OF THERMODYNAMICS

Thermodynamics, heat, closed system and open system are defined.

15.1 THE FIRST LAW OF THERMODYNAMICS
   A. The first law of thermodynamics is stated; it is another
      statement of the law of conservation of energy

15.2 THE FIRST LAW OF THERMODYNAMICS APPLIED TO SIMPLE SYSTEMS
   A. An isothermal process is defined
   B. An adiabatic process is defined
   C. An isobaric process is defined
   D. The work done in an isobaric process equals the pressure
      times the change in volume
   E. Example 15.1

15.3 HUMAN METABOLISM AND THE FIRST LAW
   A. The first law is applied to human metabolism
   B. Typical metabolic rates for a human being as a function of
      activity are listed
   C. Example 15.2

15.4 THE SECOND LAW OF THERMODYNAMICS--AN INTRODUCTION
   A. There are processes which conserve energy but never occur
   B. The Clausius statement of the second law of thermodynamics
      is given
   C. A heat engine is defined

15.5 HEAT ENGINES AND REFRIGERATORS
   A. A typical heat engine is described
   B. An internal combustion engine is described
   C. The efficiency of a heat engine is defined
   D. A Carnot engine is defined and its efficiency is expressed
      in terms of the Kelvin temperatures between which it
      operates
   E. Example 15-3
   F. The second law of thermodynamics is restated (Kelvin-Planck
      statement)
   G. A refrigerator or heat pump is defined

15.6 ENTROPY AND THE SECOND LAW OF THERMODYNAMICS
   A. The entropy change is defined
   B. Example 15-4
   C. The second law of thermodynamics is restated in terms of
      entropy

15.7 ORDER TO DISORDER
   A. The second law of thermodynamics is restated as "natural
      processes tend to move toward a state of greater disorder"
   B. This statement is applied to a number of processes such as

mixing salt and pepper, and hot and cold liquids
C. The relationship of entropy to information is discussed

15.8 UNAVAILABILITY OF ENERGY; HEAT DEATH
A. In any natural process some energy becomes unavailable to do useful work, i.e. is degraded
B. This conclusion is applied to the future evolution of the universe and leads to the conclusion that its "heat death" will occur

15.9 EVOLUTION AND GROWTH; "TIME'S ARROW"
A. The increase in entropy in the universe is applied to biological evolution; evolution does not violate the increase in entropy of the universe

15.10 STATISTICAL INTERPRETATION OF ENTROPY AND THE SECOND LAW
A. Microstates and macrostates of a system are defined
B. The number of microstates corresponding to a macrostate is tabulated for the toss of four coins
C. The percentage probability of a given macrostate is equal to the relative percentage of microstates corresponding to it
D. This is generalized to more complicated systems in which the most probable evolution of a system is towards a state of greater disorder
E. This analysis is applied to several physical systems

15.11 ENERGY RESOURCES: THERMAL POLLUTION
A. The greenhouse effect is discussed
B. Thermal pollution is discussed
C. Generating electric power from the following sources is discussed:
    fossil-fuel steam plants
    nuclear energy
    geothermal energy
    tropical seas
    hydroelectric power plants
    tidal energy
    wind power
    solar energy

VIDEO TAPE: "ENGINE OF NATURE" from The Mechanical Universe... And Beyond introduces the Carnot engine, part 1, beginning with simple steam engines.
For use with Section 15.5

VIDEO TAPE: "ENTROPY" from The Mechanical Universe... And Beyond discusses the Carnot engine, part 2, and the profound implications of entropy for the behavior of matter and the flow of time through the universe.
For use with Sections 15.6 through 15.10

NEW YORK TIMES ARTICLE: "TAPPING OCEAN'S COLD FOR CROPS AND ENERGY" describes the use of different temperature sea waters to drive two different types of electrical generators.
For use with Section 15.11

Additional problem: If electricity produced by such a power plant costs $.14 a kilowatt hour, as against $.07 a kilowatt hour generated by burning oil, what would be the percentage increase in your monthly electric bill from using an ocean-powered generator?   (Answer: it would increase by 100%)

NEW YORK TIMES ARTICLE: "OCEAN WAVES POWER A GENERATOR" describes the production of electrical energy from the kinetic energy of ocean waves.
For use with Section 15.11

Additional problem: If it takes 30 miles of ocean-front to develop 1,000 megawatts for the consumer, how much of an ocean-front would it take to supply power to a home that consumes 6 kilowatts of power?   (Answer: slightly less than one foot of shore line)

NEW YORK TIMES ARTICLE: "ELECTRICITY FROM WIND" describes a new wind powered turbine generator that will generate electricity at a cost comparable to that from fossil-fuel powered plants.
For use with Section 15.11

Additional problem: If such a generator produces 300 kilowatts of power, and the average household draws 3 kilowatts of electric power during a hot summer day, how many generators would be needed to provide the peak power requirements of a city of 100,000 households?   (Answer: 1,000 turbines)

NEW YORK TIMES ARTICLE: "BETTER WAYS TO MAKE ELECTRICITY" describes technological advances such as a new generation of gas turbines.
For use with Section 15.11

Additional problem: If gas turbines are 40% efficient and steam turbines are 33% efficient, what percentage of the fuel used to power steam turbine plants would be saved by using the newer gas turbine generators?
(Answer: about 17%)

32

CHAPTER 16

ELECTRIC CHARGE AND ELECTRIC FIELD

16.1 STATIC ELECTRICITY; ELECTRIC CHARGE AND ITS CONSERVATION
   A. The origin of the word electricity is described
   B. Static electricity and electric charge are described
   C. Unlike charges attract; like charges repel
   D. There are two types of electric charge, positive and negative
   E. The law of conservation of electric charge is stated

16.2 ELECTRIC CHARGE IN THE ATOM
   A. The atom is described and an ion is defined
   B. The process of charging an object by rubbing is explained in terms of electrons added or removed from the object, leaving it with a net charge

16.3 INSULATORS AND CONDUCTORS
   A. Conductors and insulators are defined
   B. Most metals are good conductors while most other materials are insulators
   C. This property of materials is related to the electrons in the atoms of the materials
   D. Semiconductors are discussed

16.4 INDUCED CHARGE; THE ELECTROSCOPE
   A. An electroscope is a device for detecting charge; a typical one is described

16.5 COULOMB'S LAW
   A. Coulomb's law is stated as is the unit of electric charge, the coulomb.
   B. The charge on the electron is given
   C. The permittivity of free space is defined
   D. The net force on any charge is the vector sum of the forces due to all of the other charges interacting with it

16.6 SOLVING PROBLEMS INVOLVING COULOMB'S LAW AND VECTORS
   A. A review of vector addition is provided
   B. Example 16-1
   C. Example 16-3

16.7 THE ELECTRIC FIELD
   A. The electric field is defined and an expression for it derived for a single point charge
   B. Example 16-4
   C. Example 16-5

16.8 FIELD LINES
   A. Electric field lines or lines of force are used to depict the electric field

B. These lines start on positive charges, end on negative charges, and are proportional to the magnitude of the charge
C. The electric field around point charges and between charge plates are depicted
D. The properties of electric field lines are summarized

16.9 ELECTRIC FIELDS AND CONDUCTORS
A. The electric field inside a conductor is zero when the charges are at rest
B. Any net charge on a conductor distributes itself on the outer surface
C. The electric field is always perpendicular to the surface outside of a conductor

16.10 ELECTRIC FORCES IN MOLECULAR BIOLOGY: DNA STRUCTURE AND REPLICATION
A. Molecular biology is defined
B. DNA is described: the forces holding its double helix together are electrostatic in nature
C. The replication of DNA is also due to electrostatic forces

16.11 PROTEIN SYNTHESIS AND STRUCTURE
A. Protein molecules are long chains of single small molecules known as amino acids
B. The accepted model for how amino acids are connected together in the correct order to form a protein molecule is described

VIDEO TAPE: "STATIC ELECTRICITY" from The Mechanical Universe...And Beyond introduces the principles of static electricity.
For use with Section 16.1

VIDEO TAPE: "THE ELECTRIC FIELD" from The Mechanical Universe...And Beyond discusses Michael Faraday's vision of lines of constant force in space.
For use with Section 16.7

VIDEO TAPE: PHYSICS YOU CAN SEE: "INDUCED CHARGES" demonstrates induced charges
For use with Section 16.2
Running Time: 1:10

CHAPTER 17

ELECTRIC POTENTIAL AND ELECTRIC ENERGY

17.1 ELECTRIC POTENTIAL AND POTENTIAL DIFFERENCE
   A. Electric potential is defined as is the difference in potential or potential difference
   B. The unit of electrical potential, the volt, is defined
   C. The change in potential energy of a charge moved between two points is derived
   D. Example 17-1

17.2 RELATION BETWEEN ELECTRIC POTENTIAL AND ELECTRIC FIELD
   A. The relationship of the potential difference between two points to the uniform electric field existing between the points is given
   B. Example 17-2
   C. The electric field is equal to the rate at which the electric potential changes over distance in a given direction

17.3 EQUIPOTENTIAL LINES
   A. Equipotential lines and surfaces are defined
   B. An equipotential surface must be perpendicular to the electric field at any point
   C. The equipotential lines are drawn for two oppositely charged particles

17.4 THE ELECTRON VOLT, A UNIT OF ENERGY
   A. The electron volt is defined

17.5 ELECTRIC POTENTIAL DUE TO SINGLE POINT CHARGES
   A. The potential due to a single point charge is given
   B. Example 17-3

17.6 ELECTRIC DIPOLES
   A. An electric dipole is defined
   B. The potential due to an electric dipole is derived
   C. Dipole moment is defined
   D. Polar molecules are defined
   E. Example 17-5

17.7 CAPACITANCE
   A. A capacitor is defined as is its capacitance
   B. The formula for the capacitance of a parallel plate capacitor is given
   C. Example 17-6

17.8 DIELECTRICS
   A. A dielectric is defined
   B. The effect of introducing a dielectric into a parallel plate capacitor is discussed; the permittivity of the

material introduced is described
17.9 STORAGE OF ELECTRIC ENERGY
A. An expression is derived for the energy stored in a parallel plate capacitor
B. Example 17-7
C. An expression is derived for the energy density of the electric field

17.10 THE ELECTROCARDIOGRAM
A. Each heartbeat produces changes in electrical potential that can be detected with electrodes attached to the skin
B. The record of the potential changes in a given heart is called an electrocardiogram
C. The details of the depolarization of heart cells and the resultant electrical potential at any point outside the cell are described
D. A description of the equipment used for an EKG is provided

VIDEO TAPE: "POTENTIAL AND CAPACITANCE" from The Mechanical Universe...And Beyond describes Benjamin Franklin's successful theory of how the Leyden jar works and his invention of the parallel plate capacitor.
For use with Section 17.1, 17.7, and 17.8

CHAPTER 18

ELECTRIC CURRENTS

18.1 THE ELECTRIC BATTERY
    A. The history of the discovery of the battery is described
    B. Electrodes, the electrolyte, and the electric cell are described
    C. How a simple battery works is described

18.2 ELECTRIC CURRENT
    A. Electric current is defined
    B. The ampere is defined
    C. Example 18-1

18.3 OHM'S LAW: RESISTANCE AND RESISTORS
    A. Ohm's Law is stated and discussed
    B. Resistance is defined
    C. Example 18-2

18.4 RESISTIVITY
    A. Resistivity is defined
    B. A table of resistivity and temperature coefficients is presented
    C. Example 18-3
    D. Example 18-4

18.5 SUPERCONDUCTIVITY
    A. Superconducting and the transition temperature are defined
    B. High temperature superconductors and their possible applications are discussed

18.6 ELECTRIC POWER
    A. The equation relating electric power to current and voltage is presented
    B. Example 18-5
    C. Example 18-6
    D. Fuses and circuit breakers are discussed

18.7 ALTERNATING CURRENT
    A. Direct currents and alternating currents are discussed
    B. Peak voltage and peak current are defined
    C. Average power developed by an alternating current is derived
    D. Example 18-7

18.8 NERVOUS SYSTEM AND NERVE CONDUCTION
    A. Neurons, dendrites, axons and synapse are defined and discussed
    B. The potential difference across cell membranes is described
    C. An action potential is defined and described

D. Example 18-8

VIDEO TAPE: "VOLTAGE, ENERGY AND FORCE" from The Mechanical
Universe...And Beyond describes when electricity is dangerous,
benign, spectacular, or useful
For use with Sections 18.6 and 18.7

VIDEO TAPE: "THE ELECTRIC BATTERY" from The Mechanical
Universe...And Beyond describes the invention of the electric
battery by Volta
For use with Section 18.1

VIDEO TAPE: PHYSICS YOU CAN SEE: "ELECTRIC CURRENT" describes
current conduction and why light bulbs burn out when they are
turned on
For use with Sections 18.1 and 18.3
Running Time: 3:45

CHAPTER 19

DC CURRENTS AND INSTRUMENTS

19.1 RESISTORS IN SERIES AND IN PARALLEL
   A. Resistors in series are defined
   B. Resistors in parallel are defined
   C. The single resistance equivalent to several resistances in series is determined
   D. The single resistance equivalent to several resistances in parallel is determined
   E. Example 19-1
   F. Example 19-2

19.2 EMF AND TERMINAL VOLTAGE
   A. A source of electromotive force is defined
   B. The internal resistance of a battery is defined
   C. The terminal voltage of a battery is defined
   D. The relationship between the above quantities is given
   E. Example 19-3

19.3 KIRCHHOFF'S RULES
   A. Kirchhoff's junction rule is stated
   B. Kirchhoff's loop rule is stated
   C. The sign conventions used in applying these rules are discussed

19.4 SOLVING PROBLEMS WITH KIRCHHOFF'S RULES
   A. You may choose the direction of currents arbitrarily in applying Kirchhoff's rules
   B. Example 19-4

19.5 EMF'S IN SERIES AND IN PARALLEL: CHARGING A BATTERY
   A. The resultant voltage of batteries in series and in parallel is calculated
   B. Charging a battery is described

19.6 CIRCUITS CONTAINING CAPACITORS IN SERIES AND IN PARALLEL
   A. The single capacitor equivalent to several capacitors in parallel is determined
   B. The single capacitor equivalent to several capacitors in series is determined
   C. Example 19-5

19.7 CIRCUITS CONTAINING A RESISTOR AND A CAPACITOR
   A. The time constant of such a circuit is determined
   B. The behavior of such a circuit is described

19.8 HEART PACEMAKERS
   A. An electronic pacemaker is described

19.9 ELECTRIC HAZARDS: LEAKAGE CURRENTS
   A. An electric shock can cause damage to the human body
   B. The relationship of the magnitude of electric currents to
      their physical effects are discussed
   C. The resistance of wet and dry skin are described
   D. Grounding of electric wires is discussed
   E. Leakage currents are discussed in relation to hospital
      patients with implanted electrodes

19.10 DC AMMETERS AND VOLTMETERS
   A. An ammeter and a voltmeter are defined
   B. The construction of an ammeter from a galvanometer is
      discussed
   C. The construction of a voltmeter from a galvanometer is
      discussed
   D. Multimeters and ohmmeters are described
   E. The sensitivity of a meter is discussed

19.11 CORRECTING FOR METER RESISTANCE
   A. The importance of the sensitivity of a meter is discussed
   B. Example 19-6

VIDEO TAPE: "ELECTRIC CIRCUITS" from The Mechanical Universe...
And Beyond describes the work of Wheatstone, Ohm, and
Kirchhoff and the design and analysis of how current flows
For use with Sections 19.1 through 19.5

CHAPTER 20

MAGNETISM

20.1 MAGNETS AND MAGNETIC FIELDS
    A. A magnet is described: north and south magnetic poles are defined
    B. Ferromagnetic materials are defined
    C. Magnetic fields and magnetic field lines are described
    D. The earth's magnetic field and the magnetic declination are discussed

20.2 ELECTRIC CURRENTS PRODUCE MAGNETISM
    A. An electric current produces a magnetic field
    B. The right hand rule for the magnetic field around a current carrying wire is stated

20.3 FERROMAGNETISM: DOMAINS
    A. Domains are described
    B. The Curie Temperature is defined
    C. All magnetic fields are produced by electric currents

20.4 ELECTROMAGNETS AND SOLENOIDS
    A. A solenoid is defined
    B. An electromagnet is described
    C. Applications are described

20.5 FORCE ON AN ELECTRIC CURRENT IN A MAGNETIC FIELD; DEFINITION OF B
    A. A magnet exerts a force on a current carrying wire
    B. The right hand rule for the force on a current carrying wire due to a magnetic field is described
    C. The tesla and the gauss are defined
    D. Example 20-1

20.6 FORCE ON AN ELECTRIC CHARGE MOVING IN A MAGNETIC FIELD
    A. The formula for the force on a charged particle moving in a magnetic field is given
    B. Example 20-3
    C. Example 20-4

20.7 THE HALL EFFECT
    A. The Hall effect and the Hall emf are described
    B. This effect is used to pump blood in artificial heart machines

20.8 APPLICATIONS: GALVANOMETERS, MOTORS, LOUDSPEAKERS
    A. A galvanometer is described
    B. The magnetic moment of a coil is defined
    C. A chart recorder is described
    D. An electric motor is described and its parts named
    E. A loudspeaker is described

20.9 DISCOVERY AND PROPERTIES OF THE ELECTRON
   A. Cathode rays are described
   B. Thompson's measurement of e/m of the electron is described
   C. Millikan's oil-drop experiment which determined the charge of the electron is described

20.10 THERMIONIC EMISSION AND THE CATHODE-RAY TUBE
   A. Thermionic emission is described
   B. The cathode-ray tube is described
   C. The operation of a television picture tube is described
   D. The operation of an oscilloscope is described

20.11 MASS SPECTROMETER
   A. The operation of a mass spectrometer is described
   B. Example 20-5

20.12 DETERMINATION OF MAGNETIC FIELD STRENGTHS: AMPERE'S LAW
   A. The expression for the magnetic field produced by a current carrying wire is stated
   B. Example 20-6
   C. Ampere's Law is stated
   D. The expression for the magnetic field due to a solenoid is derived
   E. Example 20-7

20.13 FORCE BETWEEN TWO PARALLEL WIRES; OPERATIONAL DEFINITION OF THE AMPERE AND THE COULOMB
   A. The expression for force per unit length between two current carrying wires is determined
   B. The ampere is defined
   C. The coulomb is defined
   D. Example 20-8

20.14 MAGNETIC FIELDS IN MAGNETIC MATERIALS; HYSTERESIS
   A. The magnetic permeability of a material is defined
   B. Saturation of a magnetic material is described
   C. A hysteresis loop is described

   VIDEO TAPE: "MAGNETISM" from The Mechanical Universe...And Beyond describes the discovery that the earth behaves like a giant magnet
   For use with 20.1

   VIDEO TAPE: "THE MILLIKAN EXPERIMENT" from The Mechanical Universe...And Beyond describes the measurement of the value of the charge of the electron
   For use with 20.9

   VIDEO TAPE: "THE MAGNETIC FIELD" from The Mechanical Universe ...And Beyond describes the law of Biot and Savart, the force between currents and Ampere's law

For use with 20.12 and 20.13

NEW YORK TIMES ARTICLE: "DIP ON EARTH IS BIG TROUBLE IN SPACE" describes the existence of a "weak spot" in the earth's magnetic field off the coast of Brazil. It adversely affects the operation of sensitive scientific equipment in orbit around the earth

Additional Problem: Suppose that the earth's magnetism were produced by a current at its geographical center. If this current moved 250 miles off-center, and the radius of the earth is about 4,000 miles, what would be the percentage change in the magnetic field observed at the equator?
(Answer: 6.7%)
For use with Sections 20.1 and 20.12

NEW YORK TIMES ARTICLE: "ROAD TO BETTER AIR FOR LOS ANGELES" describes an automobile that would run on electricity generated from the magnetic fields of cables buried in the pavement.

Additional Problem: If such a buried cable carried a current of 80 amperes, and the cable was 15 centimeters below the feet of the car's passengers, what magnetic field would their feet experience? (Answer: 1.1 X 10^{-4} Tesla)
For use with Section 20.12

NEW YORK TIMES ARTICLE: "EXPERIMENTAL PROPULSION SYSTEM HAS NO MOVING PARTS" describes a magnetohydrodynamic propulsion system for boats.

Additional Problem: What force would be generated by such a system if an electric current of 100 amperes was generated across a gap of 1.0 meters in a perpendicular magnetic field of 1.0 Tesla? (Answer: 100 Newtons)
For use with Section 20.5

CHAPTER 21

ELECTROMAGNETIC INDUCTION AND FARADAY'S LAW; AC CIRCUITS

21.1 INDUCED EMF
   A. A changing magnetic field induces an emf

21.2 FARADAY'S LAW OF INDUCTION; LENZ'S LAW
   A. The statement of Faraday's Law of Induction
   B. The statement of Lenz's Law
   C. Example 21-1

21.3 EMF INDUCED IN A MOVING CONDUCTOR
   A. The expression for the emf induced in a moving
      conductor is derived
   B. Example 21-2

21.4 CHANGING MAGNETIC FLUX PRODUCES AN ELECTRIC FIELD
   A. A changing magnetic flux produces an electric field
   B. Example 21-3

21.5 ELECTRIC GENERATORS
   A. The operation of an electric generator is described
   B. The expression for the output emf of an electric
      generator is derived
   C. Example 21-4
   D. An alternator is described

21.6 COUNTER EMF AND TORQUE; EDDY CURRENTS
   A. A back emf or counter emf is described
   B. Example 21-5
   C. Eddy currents are described

21.7 TRANSFORMERS; TRANSMISSION OF POWER
   A. Primary and secondary coils are described
   B. The transformer equation is derived
   C. Example 21-6
   D. Example 21-7

21.8 APPLICATIONS OF INDUCTION: MAGNETIC MICROPHONE,
   SEISMOGRAPH, RECORDING HEADS AND COMPUTERS
   A. The operation of a microphone is described
   B. The operation of a seismograph is described
   C. The operation of a recording head is described
   D. The storage of digital information is described

21.9 INDUCTANCE
   A. The mutual inductance constant is defined
   B. The Henry is defined
   C. The self-inductance constant is defined
   D. Impedance of an inductance is defined
   E. Example 21-8

21.10 ENERGY STORED IN A MAGNETIC FIELD
   A. The expression for the energy density in an inductor
      is stated
   B. The expression for the energy density in a magnetic field
      is derived

21.11 LR CIRCUIT
   A. An LR circuit is defined
   B. The expression for the current versus time in an LR
      circuit is given

21.12 AC CIRCUITS AND IMPEDANCE
   A. The current and voltage are in phase across a
      resistor
   B. In an inductor the current lags the voltage by a
      quarter cycle
   C. Inductive reactance or impedance is defined
   D. Example 21-9
   E. In a capacitor the current leads the voltage by one
      quarter cycle
   F. The capacitive reactance or impedance is defined
   G. Only in a resistance is energy dissipated
   H. Example 21-10

21.13 LRC SERIES AC CIRCUIT; PROBLEM SOLVING
   A. A phaser diagram is described
   B. The expression for the total impedance of an LRC
      circuit is given
   C. Electrical resonance is described

21.14 IMPEDANCE MATCHING
   A. The matching of impedances in electrical circuits is
      described
   B. The effects of impedance mismatching are described

VIDEO TAPE: PHYSICS YOU CAN SEE: "THOMPSON APPARATUS"
demonstrates electromagnetic induction.
For use with Section 21.1
Running Time: 2:29

VIDEO TAPE: "ELECTROMAGNETIC INDUCTION" from The Mechanical
Universe...And Beyond describes the discovery of
electromagnetic induction in 1831 and its use in the
generation of electric power
For use with Sections 21.1 and 21.5

VIDEO TAPE: "ALTERNATING CURRENT" from The Mechanical
Universe...And Beyond describes the generation of alternating
currents and the use of transformers to distribute electrical
currents over long distances
For use with Sections 21.1, 21.5, and 21.7

CHAPTER 22

ELECTROMAGNETIC WAVES

22.1 CHANGING ELECTRIC FIELDS PRODUCE MAGNETIC FIELDS; MAXWELL'S EQUATIONS
    A. Maxwell's equations are described verbally
    B. A changing electric field will produce a magnetic field

22.2 MAXWELL'S FOURTH EQUATION; DISPLACEMENT CURRENT
    A. Displacement current is defined
    B. Ampere's Law is extended to include displacement current

22.3 PRODUCTION OF ELECTROMAGNETIC WAVES
    A. How electromagnetic waves are produced is described
    B. The radiation field is described
    C. The electric and magnetic fields at any point are perpendicular to each other and to the direction of motion
    D. Electromagnetic waves are produced by accelerating electric charges

22.4 CALCULATION OF THE SPEED OF ELECTROMAGNETIC WAVES
    A. Plane waves are defined
    B. The speed of light is calculated to be precisely equal to the measured speed

22.5 LIGHT AS AN ELECTROMAGNETIC WAVE AND THE ELECTROMAGNETIC SPECTRUM
    A. The electromagnetic spectrum is described
    B. The relationship between the frequency, wavelength, and speed of an electromagnetic wave is stated
    C. Example 22-1
    D. Example 22-2

22.6 ENERGY IN EM WAVES
    A. Equations are derived for the energy density in any region of space at any instant
    B. An expression is derived for the rate at which energy is transported by electromagnetic waves
    C. Example 22-3

22.7 RADIO AND TELEVISION
    A. The transmission process for a radio signal is described
    B. The carrier frequency is described
    C. Amplitude modulation is described
    D. Frequency modulation is described
    E. Receivers of radio and television programs are described
    F. Example 22-4

VIDEO TAPE: "MAXWELL'S EQUATIONS" from The Mechanical Universe... And Beyond describes Maxwell's discovery that displacement current produces electromagnetic waves, or light.

For use with Sections 22.1 through 22.3

NEW YORK TIMES ARTICLE: "NEXT, DIGITAL RADIO FOR A SUPERIOR SOUND" describes the use of digital radio transmissions to produce a higher quality sound than normal radio transmissions.
For use with Section 22.7

Additional problem: If a digital radio station transmits with a power of only 1,000 watts, with a power bill of $15 a month, and is able to cover the same area as an FM station transmitting at 50,000 watts, how much money in power charges would the digital radio station save per month? (Answer: $735 per month, **not** the $6,000 per month stated in the article)

NEW YORK TIMES ARTICLE: "SCIENCE FICTION NEARS REALITY: POCKET PHONE FOR GLOBAL CALLS" describes a 25-ounce handset that would allow the user to place a call via 77 earth satellites from any point on the earth to any other point.
For use with Section 22.7

Additional problem: If Motorola uses only those microwave frequencies already set aside for sea-to-shore and air-to-ground satellite communications, what is the minimum area per simultaneous caller, on average? (Answer: 476 square miles per caller)

NEW YORK TIMES ARTICLE: "THE ELBOWING IS BECOMING FIERCE FOR SPACE ON THE RADIO SPECTRUM" describes the intense competition for the use of broadcasting frequencies.
For use with Sections 22.5 and 22.7

Additional problem: Calculate the wavelength of a 1,000 megahertz wave. (Answer: .3 meters)

NEW YORK TIMES ARTICLE: "SPACECRAFT'S IMAGES OF VENUS'S TERRAIN ASTONISH SCIENTISTS" describes the radar mapping of the surface of Venus by the Magellan spacecraft.
For use with Section 22.5

Additional problem: How small a target can be identified by radar? (Answer: it depends upon the wavelength of the radar; for example, to locate raindrops the radar waves would have to be comparable in size to raindrops.)

NEW YORK TIMES ARTICLE: "SCIENCE LEARNS TO CATCH A POLLUTING CAR IN THE ACT" describes the use of a beam of infrared light which is shined through the exhaust behind a car. The relative absorption of different frequencies by carbon monoxide and carbon dioxide allows the calculation of the amount of carbon dioxide emitted per mile.
For use with Section 22.5

Additional problem: If a "polluter" emits 250 grams of carbon monoxide per mile and travels 10,000 miles per year, what is the car's annual carbon monoxide emission?   (Answer: 2,500 kilograms of carbon monoxide per year)

NEW YORK TIMES ARTICLE: "BURSTS OF SOLAR ENERGY ENDANGER POWER LINES" describes the effects of incoming charged particles emitted by the sun upon the magnetic field that surrounds the earth and hence upon electric power systems.
For use with Sections 22.1, 22.4 and 22.6

Additional problem: If the Windsock satellite was located one million miles from earth, how long would it take a radio signal sent from the satellite to reach the earth?   (Answer: 5.4 seconds)

CHAPTER 23

LIGHT: GEOMETRIC OPTICS

23.1 THE RAY MODEL OF LIGHT
   A. The evidence supporting the ray model of light is discussed

23.2 THE SPEED OF LIGHT AND INDEX OF REFRACTION
   A. The measurements of the speed of light by Roemer and Michelson are described
   B. The speed of light in a vacuum is given
   C. The index of refraction is defined

23.3 REFLECTION; IMAGE FORMATION BY A PLANE MIRROR
   A. The law of reflection is stated
   B. The image distance and object distance are defined
   C. Real and virtual images are defined
   D. Example 23-1

23.4 FORMATION OF IMAGES BY SPHERICAL MIRRORS
   A. Convex and concave mirrors are defined
   B. The principle axis, focal point, and focal length of a mirror are defined
   C. The relationship between the radius of curvature and the focal length of a mirror is derived
   D. Spherical aberration is defined
   E. The use of ray diagrams to find the image of a given object is discussed
   F. The relation between the object distance and the image distance is derived
   G. The mirror equation is derived
   H. The lateral magnification for a curved mirror is stated
   I. Example 23-2
   J. Example 23-3
   K. Example 23-4
   L. The sign conventions for solving the mirror equation are discussed

23.5 REFRACTION; SNELL'S LAW
   A. Refraction is defined
   B. Snell's Law is stated
   C. Example 23-5

23.6 TOTAL INTERNAL REFLECTION; FIBER OPTICS
   A. The critical angle is defined
   B. Total internal reflection is described
   C. Example 23-6
   D. The use of fiber optics in medicine is described

23.7 THIN LENSES; RAY TRACING
   A. The focal point of a thin lens is described
   B. Its focal length is defined

49

    C. A converging lens and a diverging lens are defined
    D. The power of a lens is defined
    E. Ray diagraming for lenses is described

23.8 THE LENS EQUATION
    A. The lens equation is derived
    B. The sign conventions for the lens equation are stated
    C. The lateral magnification of a lens is defined

23.9 PROBLEM SOLVING FOR LENSES
    A. Example 23-7
    B. Example 23-8
    C. Example 23-9

23.10 THE LENS-MAKER'S EQUATION
    A. The lens-maker's equation is stated
    B. Example 23-11

NEW YORK TIMES ARTICLE: "NIGHT MIRROR" describes a rear-view mirror with a night position as being two reflective surfaces For use with Sections 23.3 and 23.4

See Example 23-4 for a convex rear-view car mirror.

CHAPTER 24

THE WAVE NATURE OF LIGHT

24.1 WAVES VERSUS PARTICLES; HUYGENS' PRINCIPLE AND DIFFRACTION
    A. Huygens' principle is stated
    B. Diffraction is defined

24.2 HUYGENS' PRINCIPLE AND THE LAW OF REFRACTION
    A. Huygens' principle is used to derive Snell's law of refraction
    B. The relationship to the wavelength of light in a medium and its index of refraction is stated

24.3 INTERFERENCE - YOUNG'S DOUBLE-SLIT EXPERIMENT
    A. Monochromatic light is defined
    B. Fringes on the viewing screen are described for Young's experiment
    C. The conditions for constructive interference and destructive interference are stated
    D. Example 24-1

24.4 THE VISIBLE SPECTRUM AND DISPERSION
    A. The visible spectrum is defined
    B. Dispersion is defined and described

24.5 DIFFRACTION BY A SINGLE SLIT OR DISK
    A. A diffraction pattern is described
    B. Fraunhofer diffraction and Fresnel diffraction are described
    C. The diffraction equation for the first minimum and other minima are given
    D. Example 24-3

24.6 DIFFRACTION GRADING
    A. A diffraction grading is defined
    B. The formula for the diffraction grading maxima is stated
    C. Why more lines yield sharper peaks is discussed
    D. How a diffraction grading produces a spectrum is discussed
    E. Example 24-4
    F. Example 24-5

24.7 THE SPECTROSCOPE AND SPECTROSCOPY
    A. A spectroscope is described
    B. A spectrometer is defined
    C. A line spectrum is described as is a continuous spectrum
    D. Absorption lines are described

24.8 INTERFERENCE BY THIN FILMS
    A. Interference colors from a thin oil film on water is described
    B. The phenomenon called Newton's rings is described

C. The change of phase of light reflected by a material whose index of refraction is greater than that in which it travels is discussed
D. Example 24-6
E. Lens coatings are described
F. Example 24-8

24.9 MICHELSON INTERFEROMETER
A. The construction and operation of a Michelson interferometer is described

24.10 POLARIZATION
A. A plane-polarized wave is described
B. Polarized and unpolarized light is described
C. Polaroid materials are described
D. The intensity of a plane-polarized beam of light transmitted by a polarizer is stated
E. The effects of crossed Polaroids are described
F. Example 24-9
G. Polarization by reflection is described
H. Brewster's angle is described
I. Example 24-10

24.11 OPTICAL ACTIVITY
A. Optically active substances are defined
B. Dextrorotatory and levorotatory are defined
C. The factors determining specific optical rotatory power are discussed

24.12 DOUBLE REFRACTION: BIREFRINGENCE AND DICHROISM
A. Isotropic and anisotropic materials are defined
B. The optic axis, the ordinary ray, and extraordinary ray are defined
C. Double refraction is explained
D. Dichroism is defined

24.13 SCATTERING OF LIGHT BY THE ATMOSPHERE
A. Why the sky is blue and sunsets are reddish is discussed

VIDEO TAPE: "OPTICS" from The Mechanical Universe.. And Beyond discusses many properties of light that are due to its wave aspects, including reflection, refraction, and diffraction
For use with Sections 24.1 through 24.7

NEW YORK TIMES ARTICLE: "SLEUTHS ZERO IN ON CAUSE OF TELESCOPE FLAW" discusses the error in the manufacture of the Hubble space telescope's main mirror. Experts suspect that the testing error may lie in the null corrector, which uses interference patterns to determine the curvature of the surface
For use with Sections 24.3, 24.8 and 24.9

Additional problem: How accurately can one measure distances using an interferometer? (Answer: within a quarter of the wavelength of light used. For light of wavelength 400 nanometers, this means a precision of better than 100 nanometers)

NEW YORK TIMES ARTICLE: "MEASURING DEVICE IDENTIFIED AS PROBABLE HUBBLE ERROR" indicates that apparently the light reflected off the null corrector came from the region around the pinhole rather than the metering rod behind the pinhole, introducing an error of about 1.3 millimeters
For use with Sections 24.3, 24.8 and 24.9

Additional problem: If the wavelength of light used was 400 nanometers, how many wavelengths could fit into a distance of 1.3 millimeters? (Answer: 3,200 wavelengths)

CHAPTER 25

OPTICAL INSTRUMENTS

25.1 THE CAMERA
    A. The basic elements of a camera are described
    B. Shutter speed is defined
    C. The size of the aperture opening is stated in terms of the f-stop
    D. Focusing is defined
    E. Depth of field is defined
    F. Telephoto and wide-angle lenses are described
    G. Two types of viewing systems commonly used in cameras are described
    H. Example 25-1
    I. Example 25-2

25.2 THE HUMAN EYE; CORRECTIVE LENSES
    A. The parts of the eye are described
    B. The near point and the far point of the eye are defined
    C. Nearsightedness and farsightedness are described
    D. Astigmatism is described
    E. Example 25-4

25.3 THE MAGNIFYING GLASS
    A. The properties of a simple magnifier are stated
    B. The magnifying power of the magnifier is defined
    C. Example 25-5

25.4 TELESCOPES
    A. A refracting telescope is described
    B. The formula for the magnification of the telescope is presented
    C. A reflecting telescope is described
    D. Terrestrial telescopes are described
    E. Example 25-6

25.5 COMPOUND MICROSCOPE
    A. The formula for the magnification of a microscope is derived
    B. Example 25-7

25.6 LENS ABERRATIONS
    A. Lens aberrations are defined
    B. Spherical aberration is defined
    C. Circle of least confusion is described
    D. Chromatic aberration is defined
    E. Compound lenses are discussed

25.7 LIMITS OF RESOLUTION; THE RAYLEIGH CRITERION
    A. The resolution of a lens is defined and discussed
    B. The Rayleigh criterion for the resolution of a lens is

discussed

25.8 RESOLUTION OF TELESCOPES AND MICROSCOPES
   A. The resolving power of a lens is defined
   B. Example 25-8
   C. It is not possible to resolve the detail of objects smaller
      than the wavelength of the radiation being used

25.9 RESOLUTION OF THE HUMAN EYE AND USEFUL MAGNIFICATION
   A. The factors limiting the resolution of the human eye are
      discussed
   B. The useful magnification by a microscope is limited to
      about 500x

25.10 SPECIALTY MICROSCOPES AND CONTRAST
   A. Contrast is defined
   B. The interference microscope is described
   C. The phase-contrast microscope is described

25.11 X-RAYS AND X-RAY DIFFRACTION
   A. The discovery of x-rays and their properties are described
   B. X-ray diffraction is discussed
   C. The Bragg equation is derived
   D. X-ray diffraction patterns are discussed

25.12 X-RAY IMAGING AND COMPUTERIZED TOMOGRAPHY (CAT SCANNING)
   A. Computerized tomography is described
   B. The pixel is defined and applied to the resolution
      obtainable in a CT scan

   NEW YORK TIMES ARTICLE: "POWERFUL LIGHT BEAM" describes the
   use of a sapphire cone to concentrate sunlight to a beam
   84,000 times brighter than normal sunlight found at the
   surface of the earth
   For use with Section 25.4

   Additional problem: If radiation from the sun reaches the
   earth (above the atmosphere) at a rate of about 1350 watts
   per square meter, what would the intensity of the light
   produced by this device be if it were put into orbit above
   the atmosphere? (Answer: 11×10^{7} watts per square meter)

CHAPTER 26

SPECIAL THEORY OF RELATIVITY

26.1 GALILEAN-NEWTONIAN RELATIVITY
   A. Inertial reference frames described
   B. The relativity principle states that the laws of physics are the same in all inertial reference frames
   C. Maxwell's equations did not satisfy the relativity principle
   D. In only one frame of reference was the speed of light 3.0×10^8 meters per second
   E. The ether is described

26.2 THE MICHELSON-MORLEY EXPERIMENT
   A. The Michelson-Morley experiment is described
   B. They observed no change in the speed of light due to the ether

26.3 POSTULATES OF THE SPECIAL THEORY OF RELATIVITY
   A. Einstein proposed doing away with the idea of the ether and the accompanying assumption of an absolute reference frame at rest
   B. The two postulates of special relativity are stated and discussed

26.4 SIMULTANEITY
   A. Events seen as simultaneous in one reference frame are not seen as simultaneous in another; simultaneity is relative

26.5 TIME DILATION AND THE TWIN PARADOX
   A. The time dilation formula is derived
   B. The confirmations of this formula by experiments are discussed
   C. Example 26-1
   D. Example 26-2
   E. The twin paradox is discussed

26.6 LENGTH CONTRACTION
   A. The length contraction formula is derived and discussed
   B. Example 26-3
   C. The appearance of objects moving at speeds close to that of light is discussed

26.7 FOUR-DIMENSIONAL SPACE-TIME
   A. Four-dimensional space-time continuum is described

26.8 MASS INCREASE
   A. The mass increase formula is stated
   B. Example 26-4

26.9 THE ULTIMATE SPEED
   A. The speed of light is the ultimate speed in the universe

26.10 E=MC^2; MASS AND ENERGY
   A. The formula for relativistic kinetic energy is stated
   B. E=MC^2 is stated; mass is another form of energy
   C. Example 26-5
   D. Example 26-7

26.11 RELATIVISTIC ADDITION OF VELOCITIES
   A. The formula for the relativistic addition of velocity is
      given
   B. Example 26-8

26.12 GALILEAN AND LORENTZ TRANSFORMATIONS
   A. The Galilean transformation equations are derived
   B. The Galilean velocity transformations are derived
   B. The Lorentz transformation equations are derived
   C. The relativistic velocity transformations are derived
   D. Example 26-9

26.13 THE IMPACT OF SPECIAL RELATIVITY
   A. The correspondence principle is stated and discussed

VIDEO TAPE: "THE MICHELSON-MORLEY EXPERIMENT" from The
Mechanical Universe...And Beyond describes the 1887
experiment.
For use with Section 26.2

VIDEO TAPE: "THE LORENTZ TRANSFORMATION" from The Mechanical
Universe...And Beyond describes the different measurements
for distance and time, depending upon the observer's frame
of reference
For use with Sections 26.5, 26.6, and 26.12

VIDEO TAPE: "VELOCITY AND TIME" from The Mechanical
Universe...And Beyond describes Albert Einstein's perfecting
the central ideas of physics, resulting in a new
understanding of the meaning of space and time
For use with Sections 26.3 through 26.9

VIDEO TAPE: "MASS, MOMENTUM, ENERGY" from The Mechanical
Universe...And Beyond describes the necessity to formulate
a new mechanics because of the new meaning of space and time
For use with Section 26.10

**NEW YORK TIMES ARTICLE: "DID EINSTEIN'S WIFE CONTRIBUTE TO
HIS THEORIES?"** discusses the possible role of Einstein's
first wife on the theory of relativity.
For use with Section 26.3

Additional project: Students might be asked to read the

recently published volume, edited by Dr. John Stachel,
containing Einstein's early correspondence.

CHAPTER 27

EARLY QUANTUM THEORY AND MODELS OF THE ATOM

27.1 PLANCK'S QUANTUM HYPOTHESIS
  A. Blackbody radiation is defined and described
  B. Wien's displacement law is stated
  C. Planck's quantum hypothesis is stated and the value of Planck's constant is stated
  D. A quantum of energy is defined

27.2 PHOTON THEORY OF LIGHT AND THE PHOTOELECTRIC EFFECT
  A. Photons are defined
  B. The photoelectric effect is described
  C. The work function of a metal is defined
  D. Einstein's explanation of the photoelectric effect is presented
  E. The Compton effect is described
  F. Example 27-1
  G. Example 27-2

27.3 PHOTON INTERACTIONS; PAIR PRODUCTION
  A. Pair production is defined and described
  B. The inverse of pair production occurs when an electron annihilates a positron
  C. Example 27-4

27.4 WAVE-PARTICLE DUALITY; THE PRINCIPLE OF COMPLEMENTARITY
  A. The principle of wave-particle duality is discussed
  B. The principle of complementarity is stated and discussed

27.5 WAVE NATURE OF MATTER
  A. The formula for the de Broglie wavelength of a particle is stated
  B. Example 27-5
  C. The complementarity principle applies also to matter
  D. There is a discussion of "What is an electron?"

27.6 ELECTRON MICROSCOPES
  A. The electron microscope is defined
  B. The transmission electron microscope is discussed
  C. The scanning electron microscope is discussed
  D. The scanning tunneling electron microscope is discussed

27.7 EARLY MODELS OF THE ATOM
  A. Thomson's plum pudding model is described
  B. Rutherford's experiments led him to theorize that the atom consists of a massive positive charge concentrated in a tiny nucleus, surrounded by electrons

27.8 ATOMIC SPECTRA: KEY TO THE STRUCTURE OF THE ATOM
  A. Line spectra, emission spectra and absorption spectra are

described
    B. The spectrum of hydrogen is described
    C. The Balmer series is described
    D. The Rydberg constant is given
    E. The Lyman series is described
    F. The Paschen series is described
    G. The Rutherford model of the atom was unable to explain why
       atoms emit line spectra

27.9 THE BOHR MODEL
    A. Bohr postulated that electrons could exist only in orbits
       with discrete energies and would move in them without
       radiating energy: these orbits were called stationary
       states
    B. A single photon of light was emitted in jumps between
       stationary states
    C. Bohr's quantum condition for the existence of a stationary
       state, and its quantum number, are stated
    D. The formula for the radius of the orbit of each stationary
       state is derived
    E. The formula for the energy levels in hydrogen is derived
    F. The binding energy or ionization energy is defined
    G. The Balmer series, Lyman series, and the Paschen series are
       derived from Bohr's model of the atom
    H. Example 27-7
    I. Example 27-10
    J. The correspondence principle is stated

27.10 DE BROGLIE'S HYPOTHESIS
    A. Wave-particle duality is at the root of the atomic
       structure of the atom

    VIDEO TAPE: "THE ATOM" from The Mechanical Universe...And
    Beyond describes the atom, from the ancient Greeks to the
    early twentieth century.
    For use with Sections 27.7 through 27.10

    VIDEO TAPE: "PARTICLES AND WAVES" from The Mechanical
    Universe...And Beyond describes the evidence that light can
    sometimes act as a particle
    For use with Sections 27.2, 27.4, 27.5, and 27.10

CHAPTER 28

QUANTUM MECHANICS OF ATOMS

28.1 QUANTUM MECHANICS - A NEW THEORY
    A. Quantum mechanics is defined: it deals mainly with the microscopic world of atoms and light

28.2 THE WAVE EQUATION AND ITS INTERPRETATION; THE DOUBLE-SLIT EXPERIMENT
    A. The wave function is defined
    B. The Schrodinger wave equation is described
    C. The probability of finding a particle is proportional to the square of its wave function
    D. The double slit experiment for electrons or light is described and interpreted

28.3 THE HEISENBERG UNCERTAINTY PRINCIPLE
    A. There is a limit to the accuracy of certain measurements
    B. The wave particle duality means that the act of observing produces a significant uncertainty in either the position or the momentum of an electron
    C. Heisenberg's uncertainty principle for position and momentum is stated
    D. Heisenberg's uncertainty principle for energy and time is stated
    E. Example 28-1
    F. Example 28-2
    G. Example 28-3

28.4 PHILOSOPHIC IMPLICATIONS; PROBABILITY VERSUS DETERMINISM
    A. Most physicists accept the inherently probabilistic model of nature

28.5 QUANTUM-MECHANICAL VIEW OF ATOMS
    A. Electron clouds can be interpreted as probability distributions for particles

28.6 QUANTUM MECHANICS OF THE HYDROGEN ATOM; QUANTUM NUMBERS
    A. The principal quantum number, the orbital quantum number, and the magnetic quantum number, are defined and described
    B. The Zeeman effect is described
    C. The spin quantum number is described
    D. The fine structure of energy levels is described
    E. Example 28-4
    F. The selection rule for permitted electron jumps between states is described

28.7 COMPLEX ATOMS; THE EXCLUSION PRINCIPLE
    A. The atomic number is the number of electrons in a neutral atom

B. The Pauli exclusion principle states that no two electrons can occupy the same quantum state
C. The ground-state configuration for all atoms is given in the periodic table

28.8 THE PERIODIC TABLE OF ELEMENTS
A. Shells and subshells of electrons are described
B. The periodic table is discussed in terms of the filling of these shells by electrons

28.9 X-RAY SPECTRA AND ATOMIC NUMBER
A. The line spectra of atoms are mainly due to transitions between states of the outer electron
B. Characteristic x-rays are photons emitted when an electron in a upper state drops down to fill a vacated lower state
C. A Moseley plot is discussed
D. Example 28-6
E. Bremsstrahlung is described
F. The cutoff wavelength of x-rays produced is calculated and related to the voltage across the x-ray tube producing the radiation
G. Example 28-7

28.10 FLUORESCENCE AND PHOSPHORESCENCE
A. Fluorescence is described
B. The operation of fluorescent light bulbs is described
C. Phosphorescent materials are described
D. Metastable states in atoms are described

28.11 LASERS AND HOLOGRAPHY
A. Stimulated emission of radiation is described
B. Optical pumping is described
C. Pulsed lasers and continuous lasers are described
D. Medical and other uses of lasers are described
E. Holograms are defined and described
F. A white-light hologram is described

VIDEO TAPE: "LASERDERMATOLOGY" from ABC News archives describes the use of a laser to remove birthmarks.
For use with Section 28.11
Ask students to write the American Society of Laser Medicine and Surgery, 813 Second Street, Suite 200, Wausau, Wisconsin 54401, for more information about laser surgery
Running time: 3:20

VIDEO TAPE: "THE QUANTUM MECHANICAL UNIVERSE" from The Mechanical Universe...And Beyond gives an overview on the effects of quantum mechanics on our view of the universe
For use with Sections 28.1 through 28.11

NEW YORK TIMES ARTICLE: "NO DRILL, NO NOVOCAIN IN LASER DENTAL SYSTEM" describes the use of a 3-watt laser which is

used to remove tooth and gum decay without the need for drills or scrapers.

Additional problem: What would be the special advantage of using this instrument as against using drills and scrapers? (Answer: There is no need for novocain, to which some patients are allergic)

For use with Section 28.11

NEW YORK TIMES ARTICLE: "COMPUTERIZED HOLOGRAPHY: A GIMMICK GROWS UP" describes the use of computers to generate and manipulate three-dimensional video images.

Additional problem: Ask your students how such a device might be used in the future. (Answers: to assist surgeons, to design homes and buildings, to design clothes, to project entertainment films, to illustrate strategies for professional athletic teams, and to illustrate physical principles in a physics course)

For use with Section 28.11

CHAPTER 29

MOLECULES AND SOLIDS

29.1 BONDING IN MOLECULES
   A. A chemical bond is described as are covalent bonds
   B. The bond energy or binding energy is defined
   C. Ionic bonds are described
   D. The partial ionic character of covalent bonds is discussed

29.2 POTENTIAL-ENERGY DIAGRAMS FOR MOLECULES
   A. The potential energy between two point charges is restated
   B. The potential energy function for covalent bonds is described
   C. The activation energy is defined

29.3 WEAK (VAN DER WAALS) BONDS
   A. Strong bonds and weak bonds are defined
   B. Van der Waals bonds and forces are described and discussed
   C. Weak bonds are important for understanding the activities of cells, such as DNA replication

29.4 MOLECULAR SPECTRA
   A. Additional energy levels become possible for molecules
   B. Band spectra of molecules are described
   C. Molecular rotation is described
   D. Example 29-1
   E. Molecular vibration is described
   F. The zero point energy is defined
   G. The selection rules for energy transitions with emission of a photon are described

29.5 BONDING IN SOLIDS
   A. Solid-state physics and condensed-matter physics are described
   B. Amorphous materials and those having a lattice are discussed
   C. Metallic bonds are discussed
   D. Weak bonds are discussed

29.6 BAND THEORY OF SOLIDS
   A. Bands of atomic energy levels are described
   B. The energy bands of a good conductor are described
   C. The energy bands of a good insulator are described
   D. The valence band and conduction band are defined
   E. Semiconductors are defined
   F. Holes are defined

29.7 SEMICONDUCTORS AND DOPING
   A. Doping is defined
   B. N-type and p-type semiconductors are defined and described
   C. Donor levels and acceptor levels in the energy bands are

described

29.8 SEMICONDUCTOR DIODES AND TRANSISTORS
   A. A p-n junction diode is described
   B. Forward biased and reverse biased diodes are described
   C. Example 29-2
   D. A zener diode is described
   E. A rectifier is defined
   F. Half-wave rectification and full-wave rectification are discussed
   G. A diode is called a nonlinear device
   H. A junction transistor is described
   I. An amplifier is composed of transistors

CHAPTER 30

NUCLEAR PHYSICS AND RADIOACTIVITY

30.1 STRUCTURE AND PROPERTIES OF THE NUCLEUS
   A. Protons and neutrons are defined
   B. Nucleons, nuclides, atomic number, atomic mass number
      and the neutron number are defined
   C. Isotopes are defined
   D. Natural abundances are defined
   E. The formula for the radius of a nucleus is stated
   F. The atomic mass unit is defined
   G. Nuclear spin is defined

30.2 BINDING ENERGY AND NUCLEAR FORCES
   A. Example 30-1
   B. The total binding energy of the nucleus is defined
   C. The average binding energy per nucleon is defined
   D. The strong nuclear force is described: it is a short-range
      force
   E. The weak nuclear force is described

30.3 RADIOACTIVITY
   A. The history of radioactivity is discussed
   B. Unstable isotopes occur in nature
   C. The types of radiation emitted in radioactivity are
      enumerated

30.4 ALPHA DECAY
   A. Transmutation of a daughter nucleus from a parent nucleus
      is described in terms of the emission of an alpha particle
   B. Why strong nuclear forces cannot hold a nucleus together:
      the disintegration energy is described
   C. Example 30-2

30.5 BETA DECAY
   A. Transmutation of a daughter nucleus from a parent nucleus
      is described in terms of the emission of beta ray
      (electron)
   B. The origin of this electron from within the nucleus is
      discussed
   C. Example 30-3
   D. The neutrino is hypothesized and discovered
   E. A positron is described. It is called the antiparticle to
      the electron
   F. Electron capture is described

30.6 GAMMA DECAY
   A. Gamma rays are defined and their origin discussed
   B. Metastable states and isomers are defined
   C. Internal conversion is defined

30.7 CONSERVATION OF NUCLEON NUMBER AND OTHER CONSERVATION LAWS
   A. The law of conservation of nucleon number is stated

30.8 HALF-LIFE AND RATE OF DECAY
   A. The radioactive decay law is derived
   B. The activity of a sample is stated
   C. The half-life of an isotope is defined

30.9 CALCULATIONS INVOLVING DECAY RATES AND HALF-LIFE
   A. Example 30-4
   B. Example 30-5

30.10 DECAY SERIES
   A. A decay series is defined and described

30.11 RADIOACTIVE DATING
   A. Radioactive dating can determine the age of ancient
      materials
   B. The details of the radioactive decay process are
      discussed

30.12 STABILITY AND TUNNELING
   A. The reasons that all radioactive nuclei don't decay at
      once are discussed
   B. Tunneling is described

30.13 DETECTION OF RADIATION
   A. A geiger counter is described
   B. A scintillation is described
   C. A semiconductor detector is described
   D. A photographic emulsion is described
   E. A cloud chamber is described
   F. A bubble chamber is described
   G. A spark chamber, or wire chamber, is described

CHAPTER 31

NUCLEAR ENERGY: EFFECTS AND USES OF RADIATION

31.1 NUCLEAR REACTIONS AND THE TRANSMUTATIONS OF ELEMENTS
- A. Transmutation is defined
- B. Nuclear reaction is defined
- C. In any nuclear reaction, both electric charge and nucleon number are conserved
- D. Example 31-1
- E. The Q-value is defined
- F. Example 31-2
- G. Threshold energy is defined
- H. Neutrons are the most effective projectiles for causing nuclear reactions

31.2 NUCLEAR FISSION: NUCLEAR REACTORS
- A. Nuclear fission is defined
- B. The liquid drop model of the nucleus is described
- C. A compound nucleus and fission fragments are defined
- D. A tremendous amount of energy is released in the fission of uranium 235
- E. A chain reaction is described
- F. A nuclear reactor is described
- G. Critical mass is defined
- H. Research reactors, power reactors and breeder reactors are described
- I. Risks associated with nuclear power plants are discussed
- J. The history of the development of the atomic bomb is presented
- K. Radioactive fallout is defined

31.3 FUSION
- A. Nuclear fusion is defined
- B. Example 31-4
- C. Thermonuclear bombs are described
- D. Plasma is defined
- E. Magnetic confinements of plasmas to provide thermonuclear power is discussed
- F. The Lawson criterion is discussed
- G. Inertial confinement to provide thermonuclear power is discussed

31.4 PASSAGE OF RADIATION THROUGH MATTER: RADIATION DAMAGE
- A. Ionizing radiation is defined
- B. Radiation passing through matter can do considerable damage
- C. Radiation can kill cells in biological system or turn them cancerous
- D. Somatic damage and genetic damage due to radiation are described

31.5 MEASUREMENT OF RADIATION -- DOSIMETRY
   A. The dose of radiation is defined
   B. The curie and the becquerel are defined as units of source activity
   C. Example 31-5
   D. The units of absorbed dose are defined: the roentgen, the rad and the gray
   E. The relative biological effectiveness of a given type of radiation is defined
   F. The rem and the sievert are defined
   G. The doses to which the general population and radiation workers are typically exposed are discussed
   H. Radiation film badges and radiation sickness are described
   I. Example 31-6

31.6 RADIATION THERAPY
   A. Radiation biology and nuclear medicine are defined
   B. Radiation can kill cancer cells
   C. Radiation can also be used to sterilize bandages, equipment and food

31.7 TRACERS AND IMAGING IN RESEARCH AND MEDICINE
   A. Tracers are defined
   B. Autoradiography is defined
   C. Tracers are used to determine the activity of an organ or part of an organ or to detect a tumor

31.8 EMISSION TOMOGRAPHY
   A. Single photon emission tomography is described
   B. Positive emission tomography is defined

31.9 NUCLEAR MAGNETIC RESONANCE (NMR) AND NMR IMAGING
   A. NMR is described
   B. NMR imaging is described
   C. NMR imaging produces images of great diagnostic value, both to delineate anatomy and to study metabolic processes

VIDEO TAPE: "THE GAMMA KNIFE" from ABC News archives describes the use of narrow beams of gamma radiation to destroy a tumor deep inside a child's brain.
For use with Section 31.6
Ask students why the patient has to wear a custom-tailored metal helmet. (Answer: the metal will absorb the gamma rays so that only those striking the holes in the helmet will pass through it and destroy cells in their paths)
Running Time: 3:27

VIDEO TAPE: "MAGNETIC RESONANCE IMAGING" from ABC News archives describes the use of pulsed magnetic waves to peer inside a patient's body.
For use with Section 31.9
Ask students why NMR is needed in addition to x-ray diagnostic

equipment. (Answer: NMR can highlight some soft tissues that x-rays are unable to differentiate from surrounding soft tissues)
Running Time: 4:27

NEW YORK TIMES ARTICLE: "WILL HOSPITALS BUY YET ANOTHER COSTLY TECHNOLOGY?" describes the use of a positron emission tomography unit to locate an abnormal metabolism in part of a child's brain which resulted in repeated seizures. This device uses isotopes to track metabolic rates.
For use with Section 31.8

Additional Problem: Suppose that in the future one could measure times accurately to 10^{-12} seconds, how accurately could the position of an atom emitting a positron be localized? (Answer: .03 centimeters).

NEW YORK TIMES ARTICLE: COLD FUSION STILL ESCAPES USUAL CHECKS OF SCIENCE discusses the claims of Drs. Stanley Pons and Martin Fleischmann that they had achieved nuclear fusion at room temperature using a table-top experiment.
For use with Section 31.3

Additional Problem: What conclusion about the scientific process can be drawn from the debate generated by this issue? (Answer: most scientists will remain skeptical about claims of major scientific breakthroughs until the experimental results supporting those claims have been independently duplicated)

NEW YORK TIMES ARTICLE: "NEXT BOLD STEP TOWARD FUSION PROPOSED" discusses the possible construction of the Compact Ignition Tokamak to study the ignition of fusion materials to produce short energy bursts. Supporters claim that it is important to the United States' having a leadership role in the development of fusion reactors. The cost of the machine would be about one billion dollars.
For use with Section 31.3

Additional Problem: Ask your students to suggest reasons why some scientists may oppose the building of this instrument. (Answer: the large costs may result in a reduction in funding for other scientific research areas; the payoff is far in the future; the likelihood of ultimately building controlled fusion power plants that are commercially competitive is uncertain)

NEW YORK TIMES ARTICLE: "NEW U.S. RADIOACTIVE WASTE POLICY DRAWS FIRE" discusses a new federal policy to allow increased amounts of radioactive waste to be treated as ordinary waste.
For use with Sections 31.2 and 31.5

Additional Problem: If an individual was exposed to an additional 10 mrem per day because of the new guidelines, what would be the increased dosage received over a 20-year period? (Answer: 72 rems, a very substantial dose if received at one time)

NEW YORK TIMES ARTICLE: "HIGH RADIATION DOSES SEEN FOR SOVIET ARMS WORKERS" describes the very high radioactive contamination near some Soviet nuclear weapons facilities as well as very high doses of radiation received by some arms workers. For use with Sections 31.2 and 31.5

Additional problem: How many years would it take for a typical American civilian to receive the 113 rems of radiation supposedly received by the average Soviet plutonium worker in 1951? (Answer: about 376 years)

CHAPTER 32

ELEMENTARY PARTICLES

32.1 HIGH-ENERGY PARTICLES
   A. The de Broglie wavelength of a moving particle is restated
   B. Example 32-1

32.2 PARTICLE ACCELERATORS
   A. The Van de Graff accelerator is described
   B. The cyclotron is described
   C. Example 32-2
   D. The synchrotron is described
   E. The linear accelerator is described
   F. Colliding beams are described

32.3 BEGINNINGS OF ELEMENTARY PARTICLE PHYSICS--THE YUKAWA PARTICLE
   A. A Feynman diagram pictures how a photon acts as the carrier of the electromagnetic force between two electrons
   B. Yukawa suggested the existence of a meson as the particle to mediate the strong nuclear force between two nucleons
   C. The muon is discovered
   D. The pion is discovered: it is Yukawa's meson
   E. The graviton has not yet been identified

32.4 PARTICLES AND ANTIPARTICLES
   A. Antiparticles are defined and described

32.5 PARTICLE INTERACTIONS AND CONSERVATION LAWS
   A. Baryon numbers are defined: they are conserved
   B. Lepton numbers are defined: they are conserved
   C. Example 32-3

32.6 PARTICLE CLASSIFICATION
   A. Gauge bosons are defined
   B. Leptons are defined
   C. Hadrons are defined
   D. Baryons and mesons are defined

32.7 PARTICLE STABILITY AND RESONANCE
   A. The lifetime of a given particle depends upon which force is acting
   B. A resonance is defined

32.8 STRANGE PARTICLES
   A. Strangeness is defined: it is conserved

32.9 QUARKS AND CHARM
   A. The three quarks are defined and discussed
   B. A fourth quark, and its special quantum number, called charm, are described
   C. Two additional quarks are described

32.10 THE "STANDARD MODEL"; QUANTUM CHROMODYNAMICS (QCD) AND THE ELECTROWEAK THEORY
   A. Color and flavor are defined for the 6 quarks
   B. Quantum chromodynamics is defined
   C. Gluons are defined
   D. Asymptotic freedom is defined
   E. The electroweak theory is discussed
   F. The standard model is defined

32.11 GRAND UNIFIED THEORIES
   A. The grand unified theory is defined
   B. The unification scale is defined
   C. Symmetry breaking is defined
   D. The connection to cosmology is discussed
   E. Superstring theory is defined

VIDEO TAPE: "FROM ATOMS TO QUARKS" from the Mechanical Universe...And Beyond describes electron waves which account for the periodic table and lead to the search for quarks
For use with Sections 32.3 and 32.9

NEW YORK TIMES ARTICLE: "SUPER COLLIDER'S RISING COSTS PROVOKES OPPOSITION" describes the differences within the scientific community over the construction of the Superconducting Super Collider in Texas at an estimated cost of 8 billion dollars.
For use with Section 32.2

Additional problem: What is 40 trillion electron-volts in joules? (Answer: 6.4×10^{-6} Joules)

NEW YORK TIMES ARTICLE: "NOBELS GIVEN IN PHYSICS AND CHEMISTRY" describes the award of the Nobel Prize to three scientists whose experiments had confirmed the existence of quarks.
For use with Sections 32.2 and 32.9

Additional Problem: What is the scattering experiment the three scientists performed reminiscent of? (Answer: Rutherford's scattering of a beam of alpha particles from thin sheets of metal which determined that an atom had a dense nucleus inside it: see Section 27.7)

CHAPTER 33

ASTROPHYSICS AND COSMOLOGY

ASTROPHYSICS AND COSMOLOGY ARE DEFINED

33.1 STARS AND GALAXIES
   A. The light-year is defined
   B. The galaxy is defined and described
   C. Star clusters and nebulae are defined
   D. Galaxy clusters and superclusters are defined
   E. Parallax is defined
   F. Example 33-1

33.2 STELLAR EVOLUTION: THE BIRTH AND DEATH OF STARS
   A. Absolute brightness and apparent luminosity are defined
   B. Example 33-2
   C. Apparent magnitude and absolute magnitude are defined
   D. Example 33-3
   E. Hertzsprung-Russell (HR) diagram is described
   F. Main sequence, red giants and white dwarf stars are defined
   G. Example 33-5
   H. The birth of stars from gas clouds is described
   I. Then stellar evolution leads to the main sequence
   J. Finally the star becomes a red giant
   K. Nucleosynthesis is described
   L. Depending upon the mass of the star they evolve to white
      dwarfs, neutron stars, or black holes
   M. Supernovae and pulsars are described

33.3 GENERAL RELATIVITY: GRAVITY AND THE CURVATURE OF SPACE
   A. Einstein's principle of equivalence is stated
   B. Gravitational mass is equivalent to inertial mass
   C. The curvature of a light beam near a large mass is
      described
   D. A geodesic is defined
   E. Is the universe open or closed?
   F. Gravity is described as the curvature of space-time
   G. The Schwarzschild radius is defined for a mass M
   H. A binary system is defined

33.4 THE EXPANDING UNIVERSE
   A. Why is the sky at night dark?
   B. The redshift of light is defined and supports the idea that
      the universe is expanding
   C. Hubble's law is stated
   D. Hubble's constant is given
   E. Quasars are described
   F. The cosmological principle is stated
   G. The age of the universe is estimated
   H. The steady-state model of the universe is described

33.5 THE BIG BANG AND THE COSMIC MICROWAVE BACKGROUND
   A. The Big Bang theory of the evolution of the universe is described
   B. The cosmic microwave background radiation at a temperature of 2.7K is described: its existence supports the Big Bang theory

33.6 THE STANDARD COSMOLOGICAL MODEL: THE EARLY HISTORY OF THE UNIVERSE
   A. The standard model history of the universe starts with the Big Bang
   B. The grand unified era and the hadron era are described
   C. The lepton era is next described
   D. The radiation era is next described
   E. The universe then enters the matter-dominated era
   F. The inflationary scenario is discussed

33.7 THE FUTURE OF THE UNIVERSE?
   A. Is the curvature of the universe negative, flat, or positive?
   B. The critical density of the universe and dark matter are discussed
   C. Example 33-6
   D. The deceleration parameter is described
   E. The cyclic universe theory is described
   F. The Anthropic principle is stated

ABC NEWS ARCHIVES VIDEO TAPES

CHAPTER	TITLE OF VIDEO	RUNNING TIME (MINUTES)
5	Walking in Space	3:16
12	Goodbye Gallstones	11:46
28	Laserdermatology	3:20
31	The Gamma Knife	3:27
31	Magnetic Resonance Imaging	4:27

PHYSICS YOU CAN SEE VIDEO TAPES

CHAPTER	TITLE	RUNNING TIME (MINUTES)
3	Monkey and Gun	2:07
4	Coin and Feather	1:30
5	Test Tube on a Wheel	2:40
5	Water Whirled in a Circle	1:05
7	Swivel Hips	1:24
9	Equilibrium	3:47
10	Collapse a Can	1:58
11	Singing Pipes and Singing Rod	1:25
16	Induced Charges	1:10
18	Electric Current	3:45
21	Thompson Apparatus	2:29

THE MECHANICAL UNIVERSE...AND BEYOND VIDEO TAPES

THE RUNNING TIME OF ALL VIDEO TAPES IS ABOUT 30 MINUTES

CHAPTER	TITLE OF VIDEO
1	Introduction to the Mechanical Universe
2	Vectors
2	The Law of Falling Bodies
4	Inertia
4	Newton's Laws
5	Moving in Circles
5	The Apple and the Moon
5	Kepler's Three Laws
5	The Kepler Problem
5	Fundamental Forces
6	Potential Energy
6	Conservation of Energy
7	Conservation of Momentum
8	Angular Momentum
8	Torques and Gyroscopes
11	Harmonic Motion
11	Waves
11	Resonance
13	Temperature and Gas Laws
13	Low Temperatures
15	Engine of Nature
15	Entropy

CHAPTER	TITLE OF VIDEO
16	Static Electricity
16	The Electric Field
17	Potential and Capacitance
18	Voltage, Energy, and Force
18	The Electric Battery
19	Electric Circuits
20	Magnetism
20	The Millikan Experiment
20	The Magnetic Field
21	Electromagnetic Induction
21	Alternating Current
22	Maxwell's Equations
24	Optics
26	The Michelson-Morley Experiment
26	The Lorentz Transformation
26	Velocity and Time
26	Mass, Momentum, and Energy
27	The Atom
27	Particles and Waves
28	The Quantum Mechanical Universe
33	From Atoms to Quarks

THE NEW YORK TIMES NEWSPAPER ARTICLES

CHAPTER	TITLE OF ARTICLE
4	Physics and Computers Create a Science of Aquatic Thrills
5	Finnish Scientists to Study Gravity in Eclipse
10	Sails and Wind Speed
10	Airplane Speed
10	New Plane Wing Design Greatly Cuts Drag to Save Fuel
12	Sound Over Water
12	Ocean Loudspeakers to Sound Off for Data on Global Warming
12	Designing an SST: Noise, Sonic Booms and the Ozone Layer
15	Tapping Ocean's Cold for Crops and Energy
15	Ocean Waves Power a Generator
15	Electricity from Wind
15	Better Ways to Make Electricity
20	Dip on Earth Is Big Trouble in Space
20	Road to Better Air for Los Angeles
20	Experimental Propulsion System Has No Moving Parts
22	Next, Digital Radio for a Superior Sound
22	Science Fiction Nears Reality: Pocket Phone for Global Calls

CHAPTER	TITLE OF ARTICLE
22	The Elbowing Is Becoming Fierce for Space on the Radio Spectrum
22	Spacecraft's Images of Venus's Terrain Astonish Scientists
22	Science Learns to Catch a Polluting Car in the Act
22	Bursts of Solar Energy Endanger Power Lines
23	Night Mirror
24	Sleuths Zero in on Cause of Telescope Flaw
24	Measuring Device Identified as Probable Hubble Error
25	Powerful Light Beam
26	Did Einstein's Wife Contribute to His Theories?
28	No Drill, No Novocain in Laser Dental System
28	Computerized Holography: A Gimmick Grows Up
31	Will Hospitals Buy Yet Another Costly Technology?
31	Cold Fusion Still Escapes Usual Checks of Science
31	Next Bold Step Toward Fusion Proposed
31	New U.S. Radioactive Waste Policy Draws Fire
31	High Radiation Doses Seen for Soviet Arms Workers
32	Super Collider's Rising Cost Provokes Opposition
32	Nobels Given in Physics and Chemistry

TRANSPARENCIES

The numbers below each chapter correspond to the figures in the text.

CHAPTER 3
3-4
3-9
3-10
3-12
3-17

CHAPTER 4
4-12
4-15
4-20
4-28

CHAPTER 5
5-2
5-10
5-26

CHAPTER 6
6-1

CHAPTER 7
7-9
7-13
7-16

CHAPTER 8
8-7
8-14
8-25
8-28

CHAPTER 9
9-10
9-11
9-26
9-30

CHAPTER 10
10-4
10-8
10-13
10-15
10-17

CHAPTER 11
11-2
11-3
11-5
11-17
11-20
11-28
11-35

CHAPTER 12
12-7
12-13
12-15

CHAPTER 13
13-11
13-13

CHAPTER 14
14-5

CHAPTER 15
15-6

CHAPTER 16
16-10
16-16
16-21
16-23

CHAPTER 17
17-1
17-2
17-3
17-7

CHAPTER 18
18-12
18-13

CHAPTER 19
19-3
19-6
19-7
19-12
19-13
19-17

CHAPTER 20
20-2
20-3
20-6
20-7
20-13
20-15
20-22
20-24
20-29
20-31
20-36
20-38
20-43

CHAPTER 21
21-2
21-3
21-4
21-5
21-6
21-8
21-10
21-11
21-13
21-18
21-25
21-26
21-29
21-30

CHAPTER 22
22-3
22-4
22-5
22-6
22-7
22-8
22-9
22-10
22-13
22-14

CHAPTER 23
23-7
23-8
23-11
23-12
23-13
23-14
23-19
23-26
23-28
23-29
23-30
23-31
23-33

CHAPTER 24
24-1
24-3
24-6
24-13
24-16
24-19
24-21
24-22
24-25
24-29

CHAPTER 25
25-2
25-6
25-9
25-11
25-13
25-16a
25-29

CHAPTER 26
26-2
26-3
26-7
26-8
26-14

CHAPTER 27
27-24
27-26
27-27

CHAPTER 28
28-4
28-5
28-6
28-7
28-8
28-14
28-15
28-16

CHAPTER 29
29-4
29-5
29-8
29-24
29-25
29-26
29-27

CHAPTER 30
30-1
30-8
30-9

CHAPTER 31
31-2
31-3
31-6

CHAPTER 32
32-3
32-14

CHAPTER 33
33-2
33-6
33-7
33-19
33-21
33-22